SAMPLING DISTRIBUTION AND INFERENCE

STATISTICS

Paper – II
For
S.Y.B.Sc. (ST-222)
As Per Savitribai Phule Pune University's Revised Syllabus
Effective from June 2014

Dr. P. G. DIXIT

M.Sc., M.Phil. Ph.D. (Statistics)
Vice Principal & Head of Statistics Department
Modern College, Pune - 5.

Prof. P. S. KAPRE

M.Sc., M.Phil. (Statistics)
Department of Statistics,
Modern College, Pune - 5.

Prof. V. R. PAWGI

M. Sc., M. Phil., P.G.D.C.A.
Head, Department of Statistics,
Abasaheb Garware College,
Pune – 4.

N1430

S.Y.B.Sc. STATISTICS P – II (S-II)

Fourth Edition : January 2020
         ISBN 978-93-5164-283-1

©   : **Authors**

NIRALI PRAKASHAN
Abhyudaya Pragati, 1312, Shivaji Nagar,
Off J.M. Road, PUNE – 411005
Tel - (020) 25512336/37/39, Fax - (020) 25511379
Email : niralipune@pragationline.com

➢ **DISTRIBUTION CENTRES**

PUNE

Nirali Prakashan : 119, Budhwar Peth, Jogeshwari Mandir Lane, Pune 411002, Maharashtra
(For orders within Pune)   Tel : (020) 2445 2044, Mobile : 9657703145
   Email : niralilocal@pragationline.com

Nirali Prakashan : S. No. 28/27, Dhayari, Near Asian College Pune 411041
(For orders outside Pune)   Tel : (020) 24690204 Fax : (020) 24690316; Mobile : 9657703143
   Email : bookorder@pragationline.com

MUMBAI

Nirali Prakashan : 385, S.V.P. Road, Rasdhara Co-op. Hsg. Society Ltd.,
   Girgaum, Mumbai 400004, Maharashtra; Mobile : 9320129587
   Tel : (022) 2385 6339 / 2386 9976, Fax : (022) 2386 9976
   Email : niralimumbai@pragationline.com

➢ **DISTRIBUTION BRANCHES**

JALGAON

Nirali Prakashan : 34, V. V. Golani Market, Navi Peth, Jalgaon 425001, Maharashtra,
   Tel : (0257) 222 0395, Mob : 94234 91860;
   Email : niralijalgaon@pragationline.com

KOLHAPUR

Nirali Prakashan : New Mahadvar Road, Kedar Plaza, 1st Floor Opp. IDBI Bank, Kolhapur 416 012
   Maharashtra. Mob : 9850046155; Email : niralikolhapur@pragationline.com

NAGPUR

Nirali Prakashan : Above Maratha Mandir, Shop No. 3, First Floor,
   Rani Jhanshi Square, Sitabuldi, Nagpur 440012, Maharashtra
   Tel : (0712) 254 7129; Email : niralinagpur@pragationline.com

DELHI

Nirali Prakashan : 4593/15, Basement, Agarwal Lane, Ansari Road, Daryaganj
   Near Times of India Building, New Delhi 110002 Mob : 08505972553
   Email : niralidelhi@pragationline.com

BENGALURU

Nirali Prakashan : Maitri Ground Floor, Jaya Apartments, No. 99, 6th Cross, 6th Main,
   Malleswaram, Bengaluru 560003, Karnataka; Mob : 9449043034
   Email: niralibangalore@pragationline.com

   Other Branches : Hyderabad, Chennai

Also find us on www.facebook.com/niralibooks

Statistical Thinking will one day be necessary for effective citizenship as the ability to read and write

– H.G. Wells

Preface ...

We feel indeed very happy to present this text-book of 'Statistics Paper-II' **Sampling Distributions and Inference (ST-222)** to the students of S.Y.B.Sc. The book is written according to the revised syllabus of Savitribai Phule, Pune University with effect from June, 2014.

The main purpose of the book is to provide foundation as well as comprehensive background of Probability Theory and statistical methods to beginners in simple and interesting manner. In order to make the contents of the book easier to comprehend, we have included a requisite number of illustrations, remarks, figures, diagrams etc. to elucidate statistical concepts. Application of Statistics in real life situations is emphasized through illustrative examples. Ample number of graded problems, theoretical as well as numerical are provided at the end of each chapter along with hints and answers. The numerical problems will also be useful for the F.Y.B.Sc. students computer science to prepare for Paper – III : Practicals. A list of practicals is given in the syllabus. Appendix A compiles some important mathematical results which are needed during the entire course. Values of individual terms of binomial probabilities are given in Appendix B. A specimen paper is set for student's self assessment. We have included MS-EXCEL and R-software in finding probabilities and tests of hypotheses. It is an additional feature of the book.

This book will also serve the purpose of reference book for M.B.A., C.A., M.P.M., classes.

We are thankful to Mr. D. K. Furia and the staff of Nirali Prakashan for bringing out this book in short time. Mrs. Anagha Medhekar and Mr. Santosh Bare deserve special thanks for the work done with atmost care and sincerely. Finally, our families deserve special thanks for their support, encouragement and tolerance.

We request our colleagues, teaching Statistics to offer their criticism and suggestions, for further improvement of the book.

– Authors

Kartiki Ekadashi

$Syllabus$ ... ST-222 : Sampling Distributions and Inference

1 Chi-square (χ_n^2) Distribution (10 L)

1.1   Definition χ^2 r.v. as sum of squares of i.i.d. standard normal variables, derivation of p.d.f. of χ^2 with n degrees of freedom (d.f.) using M.G.F., nature of p.d.f. curve, computations of probabilities using tables of χ^2 distribution. Mean, variance, M.G.F., C.G.F., central moments, β_1, β_2, γ_1, γ_2, mode, additive property.

1.2   Normal approximation : $\dfrac{\chi_n^2 - n}{\sqrt{2n}}$ with proof.

1.3   Distribution of $\dfrac{X}{X + Y}$ and $\dfrac{X}{Y}$, where X and Y are two independent chi-square random variables.

2. Student's t-distribution (06 L)

2.1   Definition of T r.v. with n d.f. in the form $T = \dfrac{U}{\sqrt{\chi_n^2/n}}$, where $U \to N(0, 1)$ and χ_n^2 is a χ^2 r.v. with n d.f. and U and χ_n^2 are independent r.v.s.

2.2   Derivation of p.d.f., nature of probability curve, mean, variance, moments, mode, use of tables of t-distribution for calculation of probabilities, statement of normal approximation.

3. Snedecore's F-distribution (06 L)

3.1   Definition of F r.v. with n_1 and n_2 d.f. as $F_{n_1, n_2} = \dfrac{\chi_{n_1}^2/n_1}{\chi_{n_2}^2/n_2}$ where $\chi_{n_1}^2$ and $\chi_{n_2}^2$ are independent chi-square r.v.s with n_1 and n_2 d.f. respectively.

3.2   Derivation of p.d.f., nature of probability curve, mean, variance, moments, mode.

3.3   Distribution of $1/F_{n_1, n_2}$, use of tables of F-distribution of calculation of probabilities.

3.4   Interrelations among, χ^2, t and F variates.

4. Sampling Distributions (08 L)

4.1   Random sample from a distribution as i.i.d. r.v.s. $X_1, X_2, \ldots, X_n$.

4.2   Notion of a statistic as function of $X_1, X_2, \ldots, X_n$ with illustrations.

4.3   Sampling distribution of a statistic. Distribution of sample mean $\bar{X}$ from normal, exponential and gamma distribution, Notion of a standard error of a statistic.

4.4   Distribution of $\dfrac{nS^2}{\sigma_2} = \dfrac{1}{\sigma^2} \sum_{i=1}^{n} (X_i - \bar{X})_2$ for a sample from a normal distribution using orthogonal transformation. Independence of $\bar{X}$ and S^2.

5. Exact Tests (18 L)

5.1 Tests based on chi-square distribution :

    (a) Test for independence of two attributes arranged in 2×2 contingency table. (With Yates' correction).

    (b) Test for independence of two attributes arranged in $r \times s$ contingency table, McNemar's test.

    (c) Test for 'Goodness of Fit'. (Without rounding-off the expected frequencies).

    (d) Test for $H_0 : \sigma^2 = \sigma_0^2$ against one-sided and two-sided alternatives when (i) mean is known, (ii) mean is unknown.

5.2 Tests based on t-distribution :

    (a) t-tests for population means : (i) one sample and two sample tests for one-sided and two-sided alternatives, (ii) $100 (1 - \alpha)$ % two sided confidence interval for population mean (μ) and difference of means $(\mu_1 - \mu_2)$ of two independent normal population.

    (b) Paired t-test for one-sided and two-sided alternatives.

5.3 Tests based on F-distribution :

    (a) Tests for $H_0 : \sigma_1^2 = \sigma_2^2$ against one-sided and two-sided alternatives when (i) means are known, (ii) means are unknown.

•••

LIST OF PRACTICE

Sr. No.	Title of the experiment	No. of practicals
1.	Fitting of negative binomial distribution, testing goodness of fit.	1
2.	Fitting of normal distribution, testing goodness of fit (also using qq-plot)	1
3.	Applications of normal, negative binomial and multinomial distribution	2
4.	Model sampling from exponential, normal distribution using (i) distribution function, (ii) Box- Muller transformation.	1
5.	Time series : Estimation and forecasting of trend by fitting of AR (1) model, exponential smoothing, moving averages.	1
6.	Estimation of seasonal indices by ratio to trend.	1
7.	Test for means and construction of confidence interval. (Also using MS-EXCEL) (i) $H_0 : \mu = \mu_0$, σ^2 known and σ^2 unknown (ii) $H_0 : \mu_1 = \mu_2$, σ_1, σ_2 known (iii) $H_0 : \mu_1 = \mu_0$, $\sigma_1 = \sigma_2 = \sigma$ unknown (iv) $H_0 : \mu_1 = \mu_2$, paired t test	2
8.	Tests for proportions and construction of confidence interval for $P = P_1 - P_2$. $H_0 : P = P_0$, $H_0 : P_1 = P_2$	1
9.	Tests based on χ^2 distribution (i) Goodness of fit. (ii) Independence of attributes (2×2, $m \times n$ contingency table), (iii) Mc Nemar's test. (iv) $H_0 : \sigma^2 = \sigma_0^2$, μ unknown, confidence interval for σ^2	2
10.	Tests based on F-distribution $H_0 : \sigma_1^2 = \sigma_2^2$. (i) means known, (ii) means unknown.	1
11.	Fitting of multiple regression plane using MS-EXCEL.	1
12.	Fitting of normal distribution using MS-EXCEL.	1
13.	Exponential smoothing using MS-EXCEL.	1
14.	Computations of probabilities of Normal, Exponential gamma χ^2, t, F using R.	1
15.	Use of basic R software commands, finding summary statistics using R software	1
16.	Tests using R software	1
17.	**Project :** Project based on analysis of data collected by students in groups of maximum 6 students. (Project is equivalent to five practicals)	5

●●●

Contents ...

•••

Chapter 1 ...
Chi-Square (χ^2) Distribution

Yates

Born on May 12 1902 in Manchester England he was the eldest of five children. He began his mathematic education at a private school were he was encouraged and tought by an advanced mathematics teacher. Yates recieved a scholarship to Clifton College and was later awarded another scholarship at St. Johns College in Cambridge. He taught mathematics for a short period of time but wanted to use mathematics to help others instead of just teaching it. A few years later he went to Africa where he was part of the gold coast survey. A few years later he moved back and was appointed assistant statistician at Rothamsted Experimental Station. When a friend, R.A Fisher became chairman at the University College London in 1933, Yates was appointed the Head of Statistics at Rothamsted. He held that postition until 1968 when he retired. During the time he was working, he worked often with Fisher on experimental design, block designs, theory of variance and operational research.

Contents ...

Key Words :

Gamma distribution, chi-square distribution, degrees of freedom (d.f.), normal approximation.

Objectives :

(1) To understand chi-square distribution as a particular case of gamma distribution.

(2) To study the relationship between chi-square distribution and other distributions. Also limiting behaviour of chi-square distribution as $n \to \infty$ (normal approximation).

(3) Computation of probabilities using χ^2 tables and also using MS-EXCEL, R-software.

(4) To study the properties of χ^2 distribution.

1.0 Introduction

The gamma distribution is related with the theory of independent normal variates in natural way. Because we have shown that if $X_1, X_2, \ldots X_i, \ldots, X_n$ are n independent standard normal variates then the distribution of $\sum_{i=1}^{n} X_i^2$ has $G\left(\dfrac{1}{2}, \dfrac{n}{2}\right)$ distribution. This particular case of gamma distribution is called as *chi-square* distribution with n degrees of freedom (d.f.). The random variable involved in this distribution is denoted as χ_n^2. The degrees of freedom (d.f.) represents the number of independent variables used in the construction of χ^2. Thus χ_n^2 variate can be looked-upon as the sum of squares of n independent standard normal variates.

Tests based on chi-square distribution like chi-square test of goodness of fit, chi-square test of independence of attributes etc. are widely used. It also plays an important role in statistical inference.

1.1 Derivation of P.D.F. of χ^2 with n Degrees of Freedom (using M.G.F.)

Let $X_1, X_2, \ldots X_i, \ldots, X_n$ be i.i.d. N (0, 1) variates. Then m.g.f. of X_i^2 is given by,

$$M_{X^2}(t) = \int_{-\infty}^{\infty} e^{tx^2} f(x)\, dx \quad \text{where, } f(x) \text{ is p.d.f. of standard normal distribution.}$$

$$\therefore M_{X^2}(t) = \int_{-\infty}^{\infty} e^{tx^2} \cdot \frac{1}{\sqrt{2\pi}}\, e^{-\frac{1}{2}x^2}\, dx$$

$$= \frac{2}{\sqrt{2\pi}} \int_{0}^{\infty} e^{-\left(\frac{1}{2}-t\right)x^2}\, dx \left[\because \text{Integrand is even function of x, Also } a = \frac{1}{2}-t, b = 1\right]$$

$$= \frac{2}{\sqrt{2\pi}} \cdot \frac{1}{2} \frac{\Gamma\left(\frac{1}{2}\right)}{\left(\frac{1}{2}-t\right)^{1/2}} \quad ; \quad \frac{1}{2}-t > 0$$

$$= \frac{2}{\sqrt{2\pi}} \cdot \frac{1}{2} \frac{\Gamma\left(\frac{1}{2}\right)}{(1-2t)^{1/2}} \, 2^{1/2} = \left(\frac{1}{1-2t}\right)^{1/2}$$

$$= (1-2t)^{-1/2}$$

$$\therefore \qquad M_{X^2}(t) = \left(1 - \frac{t}{1/2}\right)^{\frac{-1}{2}} \qquad ; \qquad t < \frac{1}{2}$$

$$\because \;\; \text{If } X_i \text{ is N (0, 1) variate } M_i^2(t) = \left(1 - \frac{t}{1/2}\right)^{-\frac{1}{2}} \qquad ; \qquad t < \frac{1}{2}$$

Suppose, $Y = \sum\limits_{i=1}^{n} X_i^2$. Then Y has χ^2 distribution with n d.f. $\left(\chi_n^2\right)$. The m.g.f. of Y is as follows :

$$M_Y(t) \;\; = \;\; M_{\sum\limits_{i=1}^{n} X_i^2}(t)$$

$$= \;\; M_{X_1^2}(t) . M_{X_2^2}(t) \ldots M_{X_i^2}(t) \ldots M_{X_n^2}(t)$$

$$= \left(1 - \frac{t}{1/2}\right)^{\frac{-n}{2}} \qquad\qquad\qquad \left[\because X_i\text{'s are independent}\right]$$

It is m.g.f. of $G\left(\frac{1}{2}, \frac{n}{2}\right)$. Thus $\sum\limits_{i=1}^{n} X_i^2 = Y$ has $G\left(\frac{1}{2}, \frac{n}{2}\right)$. Hence p.d.f. Y is given as,

$$f(y) \;\; = \;\; \frac{\left(\frac{1}{2}\right)^{\frac{n}{2}}}{\Gamma\left(\frac{n}{2}\right)} \, e^{\frac{-y}{2}} \, y^{\frac{n}{2}-1} \qquad ; \qquad y \geq 0$$

$$= 0 \qquad\qquad\qquad\qquad ; \qquad \text{otherwise}$$

$$\text{OR} \qquad f(y) \;\; = \;\; \frac{1}{(\sqrt{2})^n \, \Gamma\left(\frac{n}{2}\right)} \, e^{\frac{-y}{2}} \, y^{\frac{n}{2}-1} \qquad ; \qquad y \geq 0 \qquad\qquad \ldots (1)$$

Thus a continuous variable Y is said to follow chi-square distribution with n d.f. if its p.d.f. is given by relation (1).

Another method :

Let $X_1, X_2, \ldots, X_i \ldots X_n$ be i.i.d. N(0, 1) variates.

Suppose $Y_i = X_i^2$ we shall obtain distribution function of Y. It is given by

$$G_Y(y) \;\; = \;\; P[Y \leq y] = P[X_i^2 \leq y]$$

$$= \;\; P(-\sqrt{y} \leq X_i < \sqrt{y})$$

$$= \;\; F(\sqrt{y}) - F(-\sqrt{y}) \qquad\qquad [F(\cdot) \text{ is distribution function of } X_i]$$

In order to obtain p.d.f. of y, we take derivatives of both sides w.r.t. y,

$$g(y) = \frac{d\,G(y)}{dy} = \frac{d}{dy}\,[F(\sqrt{y}) - f(-\sqrt{y})]$$

$$= \frac{f(\sqrt{y})}{2\sqrt{y}} - \frac{f(-\sqrt{y})}{2\sqrt{y}}(-1)$$

$$= \frac{f(\sqrt{y}) + f(-\sqrt{y})}{2\sqrt{y}} \qquad [\because X \to N(0, 1),\ f(x) = \frac{1}{\sqrt{2\pi}}\,e^{-x^2/2}]$$

$$= \frac{\dfrac{1}{\sqrt{2\pi}}\,e^{-y/2} + \dfrac{1}{\sqrt{2\pi}}\,e^{-y/2}}{2\sqrt{y}}$$

$$\therefore \qquad g(y) = \frac{\dfrac{1}{\sqrt{2\pi}}\,e^{-y/2}}{\sqrt{y}} \quad ; \quad y > 0$$

$$= \frac{\left(\dfrac{1}{2}\right)^{1/2}}{\Gamma\left(\dfrac{1}{2}\right)}\,e^{-y/2}\,y^{\frac{1}{2}-1} \ ; \ y > 0 \qquad [\because \Gamma\tfrac{1}{2} = \sqrt{\pi}]$$

which p.d.f. of gamma distribution with parameters $\left(\dfrac{1}{2},\dfrac{1}{2}\right)$. Thus X_i^2 follows $G\left(\dfrac{1}{2},\dfrac{1}{2}\right)$ distribution i = 1, 2 … n.

$\because$ Each X_i^2 follows $G\left(\dfrac{1}{2},\dfrac{1}{2}\right)$ distribution. Using additive property of gamma variates.

$T = \displaystyle\sum_{i=1}^{n} X_i^2$ follows $G\left(\dfrac{1}{2},\dfrac{n}{2}\right)$ distribution. The p.d.f. of T which has χ_n^2 probability distribution is given by

$$f(t) = \frac{1}{(\sqrt{2})^n\,\Gamma\left(\dfrac{n}{2}\right)}\,e^{-t/2}\,t^{n/2-1} \ ; \ t \geq 0$$

Remarks :

1. The probability density curves for n = 1, 2, 3, 4, 5, and 6 are as shown in figure 1.1.

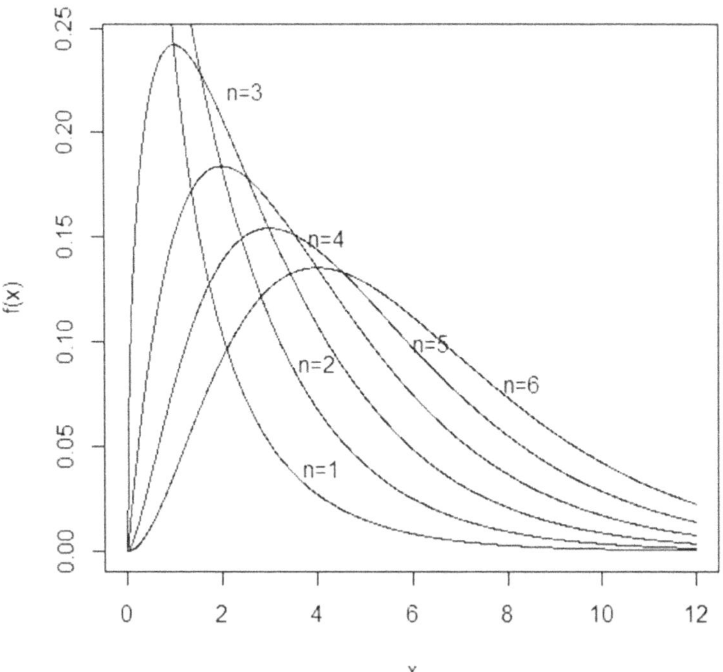

Fig. 1.1 : Chi-square probability density

The probability density curve of χ^2 distribution has longer tail on right hand side. Thus the probability density curve is positively skew. It is also clear from Fig. 1.1 that mode of chi-square distribution exists only when n > 2.

2. Some practical situations in which χ^2 variate occurs can also be stated. For instance, suppose X represents the   error in measuring radius of a circle and $X \to N\ (0, 1)$. Then error in measuring area of the circle = $A = \pi X^2$

$\therefore\ \ Y = \dfrac{A}{\pi} = X^2$ follows chi-square distribution with 1 d.f.

1.2 Moments of Chi-square Distribution

If $Y \to \chi^2_n$, then r^{th} raw moment of Y is given by,

$$\mu'_r = E\left(Y^r\right) = \int_0^\infty y^r\, f(y)\, dy = \frac{\left(\frac{1}{2}\right)^{n/2}}{\Gamma\left(\frac{n}{2}\right)} \int_0^\infty y^r\, e^{\frac{-y}{2}}\, y^{\frac{n}{2}-1}\, dy$$

$$= \frac{\left(\frac{1}{2}\right)^{n/2}}{\Gamma\left(\frac{n}{2}\right)} \int_0^\infty e^{\frac{-y}{2}}\, y^{\frac{n}{2}+r-1}\, dy = \frac{\left(\frac{1}{2}\right)^{n/2}}{\Gamma\left(\frac{n}{2}\right)} \cdot \frac{\Gamma\left(\frac{n}{2}+r\right)}{\left(\frac{1}{2}\right)^{\frac{n}{2}+r}}$$

$$\therefore \qquad \boxed{\mu'_r = \frac{\Gamma\left(\frac{n}{2} + r\right)}{\Gamma\left(\frac{n}{2}\right)} \cdot 2^r} \qquad\qquad \text{... (i)}$$

$$= 2^r \left(\frac{n}{2} + r - 1\right)\left(\frac{n}{2} + r - 2\right) \cdots \left(\frac{n}{2} + 1\right)\left(\frac{n}{2}\right)$$

$$\therefore \qquad \boxed{\mu'_r = n\,(n + 2) \cdots (n + 2r - 2);\ r = 1, 2, 3 \ldots} \qquad\qquad \text{... (ii)}$$

putting r = 1, 2, 3 and 4 in equation (ii) we get first four raw moments as;

$$\mu'_1 = \text{Mean} = n$$

$$\mu'_2 = n(n + 2)$$

$$\mu'_3 = n(n + 2)\,(n + 4)$$

$$\mu'_4 = n(n + 2)\,(n + 4)\,(n + 6)$$

Remarks :

1. From above formulae we have

$$\text{Var (Y)} = \mu'_2 - \left(\mu'_1\right)^2$$

$$= n(n + 2) - n^2 = 2n$$

It is convenient to use cumulant generating function (c.g.f.) to obtain central moments.

2. Recurrence relation between raw moments can be obtained from equation (i). Replacing r by (r – 1) in equation (i) we get,

$$\mu'_{r-1} = \frac{\Gamma\left(\frac{n}{2} + r - 1\right)}{\Gamma\left(\frac{n}{2}\right)} \cdot 2^{r-1}$$

$$\frac{\mu'_r}{\mu'_{r-1}} = \left(\frac{n}{2} + r - 1\right) \cdot 2$$

$$\therefore \qquad \mu'_r = 2\left(\frac{n}{2} + r - 1\right) \cdot \mu'_{r-1}\,;\ r = 1, 2, 3, \ldots$$

For example : Putting r = 1, $\mu'_1 = n$, r = 2, $\mu'_2 = 2\left(\frac{n}{2} + r - 1\right)\mu'_1 = (n + 2)\,n$ etc.

M.G.F. and C. G. F. of χ^2 distribution

Let $Y \to \chi_n^2$. Then m.g.f. of Y is given by

$$M_Y(t) = E[e^{ty}]$$

$$= \frac{\left(\frac{1}{2}\right)^{\frac{n}{2}}}{\Gamma\left(\frac{n}{2}\right)} \int_0^\infty e^{ty} \, e^{\frac{-y}{2}} \, y^{\frac{n}{2}-1} \, dy$$

$$= \frac{\left(\frac{1}{2}\right)^{\frac{n}{2}}}{\Gamma\left(\frac{n}{2}\right)} \int_0^\infty e^{-\left(\frac{1}{2}-t\right)y} \, y^{\frac{n}{2}-1} \, dy, \qquad \text{if } \left(\frac{1}{2}-t\right) > 0$$

$$= \frac{\left(\frac{1}{2}\right)^{\frac{n}{2}}}{\Gamma\left(\frac{n}{2}\right)} \cdot \frac{\Gamma\left(\frac{n}{2}\right)}{(1-2t)^{n/2}} \cdot 2^{\frac{n}{2}}, \qquad \left(\text{if } t < \frac{1}{2}\right)$$

$$\boxed{M_Y(t) = (1-2t)^{-\frac{n}{2}}}, \qquad \left(\text{if } t < \frac{1}{2}\right)$$

In order to obtain raw moments, we expand R.H.S. of above equation in terms of powers of t.

$$\therefore \quad M_Y(t) = 1 - \left(\frac{-n}{2}\right)(2t) + \frac{\left(\frac{-n}{2}\right)\left(\frac{-n}{2}-1\right)}{2!} 2^2 t^2$$

$$+ \dots + \frac{(-1)^r\left(-\frac{n}{2}\right)\left(-\frac{n}{2}-1\right)\dots\dots\left(-\frac{n}{2}-r+1\right)}{r!} 2^r t^r + \dots$$

$$\therefore \quad \mu_r' = \text{coefficient of } \frac{t^r}{r!} = n(n+2) \dots (n+2r-2)$$

Note : Putting r = 1, 2, 3, 4 the above expression we get first four raw moments as

$$\mu_1' = n, \quad \mu_2' = n(n+2), \quad \mu_3' = (n+2)(n+4) \text{ and } \mu_4' = n(n+2)(n+4)(n+6)$$

The cumulant generating function is given by

$$K_Y(t) = \ln M_Y(t)$$

$$= -\frac{n}{2} \ln(1-2t)$$

$$= \frac{n}{2}\left\{2t + \frac{2^2 t^2}{2} + \frac{2^3}{3}t^3 + \frac{2^4 t^4}{4} + \dots + \frac{2^r t^r}{r} + \dots\right\}$$

$\therefore$ First cumulant = k_1 = Coefficient of $\dfrac{t^1}{1!}$ = n

Second cumulant = k_2 = Coefficient of $\dfrac{t^2}{2!}$ = 2n

Third cumulant = k_3 = Coefficient of $\dfrac{t^3}{3!}$ = 8 n

Fourth cumulant = k_4 = Coefficient of $\dfrac{t^4}{4!}$ = $(3)2^4$ n = 48 n

In general r^{th} cumulant = k_r = coefficient of $\dfrac{t^r}{r!}$

$$= n(r-1)\,!\;2^{r-1} \qquad ; \qquad r = 1, 2 \ldots$$
$$= (r-1)\,!\;n\;2^{r-1} \qquad ; \qquad r = 1, 2 \ldots$$

Remark : Then the first four central moments are obtained as follows :

$$\mu_1 = 0, \qquad \mu_2 = k_2 = 2n, \qquad \mu_3 = k_3 = 8n$$
$$\mu_4 = k_4 + 3k_2^2 = 48\,n + 12\,n^2$$

Coefficients of Skewness and Kurtosis

The coefficients of skewness β_1 and γ_1 are given by,

$$\beta_1 = \frac{\mu_3^2}{\mu_2^3} = \frac{(8n)^2}{(2n)^3} = \frac{8}{n};$$

$$\gamma_1 = \sqrt{\beta_1} = \sqrt{\frac{8}{n}}$$

Hence as $n \to \infty$, $\beta_1 \to 0$ and $\gamma_1 \to 0$. It means that when d.f is very large, the distribution becomes approximately symmetric.

The coefficients of Kurtosis β_2 and γ_2 are,

$$\beta_2 = \frac{\mu_4}{\mu_2^2} = \frac{48\,n + 12\,n^2}{4n^2} = 3 + \frac{12}{n}$$

$$\text{and} \quad \gamma_2 = \beta_2 - 3 = \frac{12}{n}$$

Therefore as $n \to \infty$, $\beta_2 \to 3$ and $\gamma_2 \to 0$.

Thus distribution becomes mesokurtic for large d.f. (n).

1.3 Recurrence Relation Among the Central Moments

Let Y follow χ^2 distribution with n. d.f.

Then the m.g.f. is given by,

$$M_Y(t) = (1 - 2t)^{-\frac{n}{2}}$$

Hence the m.g.f. for obtaining central moments.

(Central m.g.f.) is given by, $M_{Y - E(Y)}(t) = M_{Y - n}(t) = e^{-nt} M_Y(t)$

$$M_{Y-n}(t) = e^{-nt}(1 - 2t)^{\frac{-n}{2}} \qquad \because E(Y) = n$$

Taking logarithms of both sides

$$ln\, M_{Y-n}(t) = -nt - \frac{n}{2} ln(1 - 2t)$$

Differentiating both sides w.r.t. t we get

$$\frac{M'_{Y-n}(t)}{M_{Y-n}(t)} = -n + \frac{n}{2} \times \frac{2}{1 - 2t}$$

$$= \frac{2nt}{1 - 2t}$$

$$\therefore \quad (1 - 2t)\, M'_{Y-n}(t) = 2nt\, M_{Y-n}(t)$$

Differentiating both sides w.r.t. t, r times

$$(1 - 2t)\, M^{r+1}_{Y-n}(t) - 2r\, M^{r}_{Y-n}(t) + \ldots\ldots$$

$$= 2n\left[t\, M^{r}_{Y-n}(t) + r\, M^{r-1}_{Y-n}(t) + \ldots\ldots\right]$$

where, $\qquad M^r(t) = \dfrac{d^r\, M(t)}{d\, tr}$

$$= r^{th} \text{ derivative of } M(t) \text{ w.r.t. } t.$$

Putting $\quad t = 0$, we have

$$\mu_{r+1} - 2r\, \mu_r = 0 + 2n\, r\, \mu_{r-1}$$

$\therefore \qquad \boxed{\mu_{r+1} = 2r\,[n\,\mu_{r-1} + \mu_r] \text{ for } r = 1, 2, 3\ldots}$

Note : The above recurrence relation can be directly proved using integrals.

Deductions : $\mu_0 = 1$, $\mu_1 = 0$

$\therefore$ Putting r = 1 in above relation $\mu_2 = 2\,[n\mu_0 + \mu_1] = 2n$

Putting r = 2 and r = 3 we obtain μ_3 and μ_4 as

$$\mu_3 = 8n, \qquad \mu_4 = 48\,n + 12\,n^2$$

1.4 Additive Property

Statement : If Y_1 and Y_2 are independent χ^2 variates with n_1 and n_2 d.f. respectively, then $Y_1 + Y_2$ has also χ^2 distribution with $(n_1 + n_2)$ degrees of freedom.

Proof : $Y_1 \to \chi^2_{n_1}$

$\therefore$ The m.g.f. of $Y_1 = M_{Y_1}(t) = (1 - 2t)^{-n_1/2}$

Similarly, $Y_2 \to \chi^2_{n_2}$

$\therefore$ The m.g.f. of $Y_2 = M_{Y_2}(t) = (1 - 2t)^{-n_2/2}$

The m.g.f. of $Y_1 + Y_2$ is given by

$$M_{Y_1 + Y_2}(t) = M_{Y_1}(t) \cdot M_{Y_2}(t) \qquad (\because Y_1 \text{ and } Y_2 \text{ are independent})$$

$$= (1 - 2t)^{-n_1/2} (1 - 2t)^{-n_2/2}$$

$$= (1 - 2t)^{-\left(\frac{n_1 + n_2}{2}\right)}$$

which is m.g.f. of χ^2 distribution with $n_1 + n_2$ d.f.

Thus $Y_1 + Y_2 \to \chi^2_{n_1 + n_2}$

Remark : Generalisation of the above stated property holds good. Hence if $Y_1, Y_2, ... Y_i, ..., Y_k$ are independent χ^2 variates with $n_1, n_2, ... n_i, ... n_k$ d.f. respectively then $\sum\limits_{i=1}^{k} Y_i$ follows χ^2 distribution with $\sum\limits_{i=1}^{k} n_i$ d.f.

1.5 Mode of the Chi-square Distribution

The value of y which maximizes f(y) is the mode of the distribution. Hence, mode will be the value such that $f'(y) = 0$ and $f''(y) < 0$.

Let Y be a χ^2 variate with n. d.f.

Then the p.d.f. is given by,

$$f(y) = \frac{\left(\frac{1}{2}\right)^{n/2}}{\Gamma\left(\frac{n}{2}\right)} e^{-y/2} \, y^{\frac{n}{2} - 1} \quad ; \quad y > 0$$

$$\therefore \quad \log_e f(y) = \log_e c - \frac{y}{2} + \left(\frac{n-2}{2}\right) \log_e y$$

$$\text{where, } c = \left[\frac{\left(\frac{1}{2}\right)^{n/2}}{\Gamma\left(\frac{n}{2}\right)}\right]$$

Differentiating both sides w.r.t. y we have

$$\frac{f'(y)}{f(y)} = 0 - \frac{1}{2} + \frac{n-2}{2y}$$

$$\therefore \quad f'(y) = \frac{f(y)}{2}\left[\frac{n-2}{y} - 1\right]$$

$$\therefore \quad f'(y) = 0 \quad \frac{f(y)}{2}\left[\frac{n-2}{y} - 1\right] = 0$$

$$\therefore \quad y = n - 2 \qquad \qquad \text{... (i)}$$

$$f''(y) = \frac{f'(y)}{2}\left[\frac{n-2}{y} - 1\right] + \frac{f(y)}{2}\left[-\frac{(n-2)}{y^2}\right]$$

$$\left[f''(y)\right]_{y = n-2} = 0 - \frac{f(y)(n-2)}{2y^2} < 0 \qquad \text{for n > 2 ... (ii)}$$

From equations (i) and (ii) mode the distribution is n – 2.

Note : From Fig. 1.1, we can get the idea that mode of χ_n^2 exists for n > 2.

1.6 (A) Use of χ^2 Tables for Calculation of Probabilities

Let Y be a χ^2 variate with n d.f. We need to find the probabilities such as P[Y ≥ a], P[Y ≤ b], P [c ≤ y ≤ d] where a, b, c, d are constants. The computations of these probabilities by evaluating the integral is a very difficult task. However, the computations of these probabilities is facilitated by using the tables giving values of χ^2 variates for various d.f. from 1 to 70. [Table of distribution of χ^2 from statistical table].

Suppose $\chi_{n,\,\alpha}^2$ represents the value of χ^2 variate such that

$$P\left[Y \geq \chi_{n,\,\alpha}^2\right] = \int_{\chi_{n,\,\alpha}^2}^{\infty} \frac{1}{2^n\,\Gamma\left(\frac{n}{2}\right)} y^{n/2}\, e^{-y/2}\, dy = \alpha$$

Then values of $\chi_{n,\,\alpha}^2$ are available in the table for α = 0.99, 0.98, 0.95 , 0.90, 0.80, 0.70, 0.50, 0.30, 0.20, 0.10, 0.05, 0.2, 0.01 and 0.001. The area corresponding to this probability is shaded area as shown in Fig. 1.2.

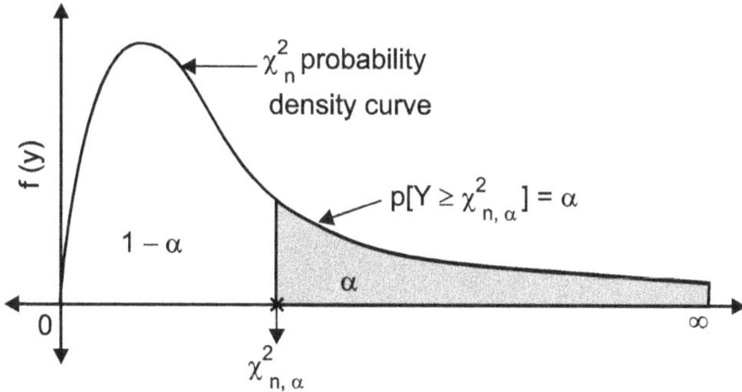

Fig. 1.2

Illustration : Let $Y = \chi^2_{11}$. Find : (i) $P[Y > 14.631]$ (ii) $P[Y \leq 8.148]$ (iii) $P[4.575 < Y < 17.275]$.
(iv) Median of distribution of Y,   (v) Second and seventh deciles of distribution of Y.

Solution :

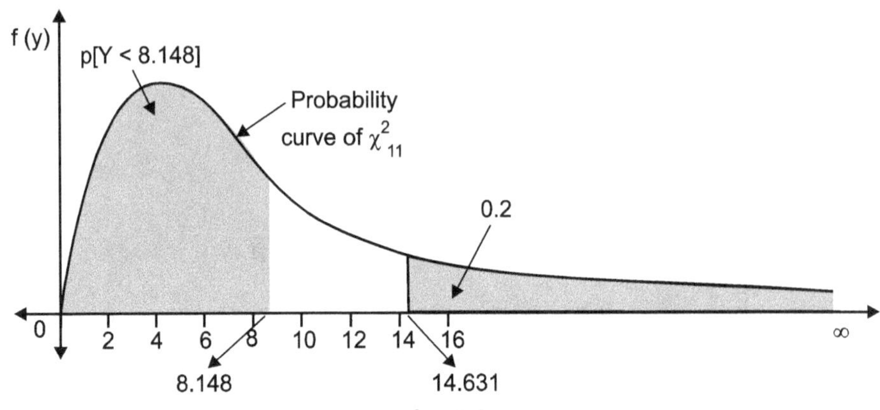

Fig. 1.3

(i) $P[Y > 14.631] = 0.2$ [From statistical tables]

(ii) $P[Y \leq 8.148]$ = $P\left[\chi^2_{11} \leq 8.148\right]$ = $1 - P\left[\chi^2_{11} > 8.148\right]$

$= 1 - 0.7 = 0.3$

(iii) $P[4.575 < Y < 17.275]$

$= P\left[4.575 < \chi^2_{11} < 17.275\right] = P\left[\chi^2_{11} \geq 4.575\right] - P\left[\chi^2_{11} \geq 17.275\right]$

$= 0.95 - 0.1 = 0.85$

(iv) Suppose median = M, then $P[Y \geq M] = 0.5$

$\therefore \quad P\left[\chi^2_{11} \geq M\right] = 0.5$ As shown in Fig. 1.4.

$P\left[\chi^2_{11} \geq 10.341\right] = 0.5$

$\therefore \qquad M = 10.341$

(v) Let D_2 and D_7 represent second and seventh deciles respectively then
$P[Y \leq D_2] = 0.2$ $P[Y > D_2] = 0.8$

$\therefore \quad P[Y > D_2] = 0.8$ As shown in Fig. 1.5,

$P\left[\chi^2 > 6.989\right] = 0.8$

$\therefore \qquad D_2 = 6.989$

Similarly, $P(Y \leq D_7) = 0.7$

$P(Y > D_7) = 0.3$

From Fig. 1.5, $P\left[\chi^2_{11} > 12.899\right] = 0.3$

$\therefore \qquad D_7 = 12.899$

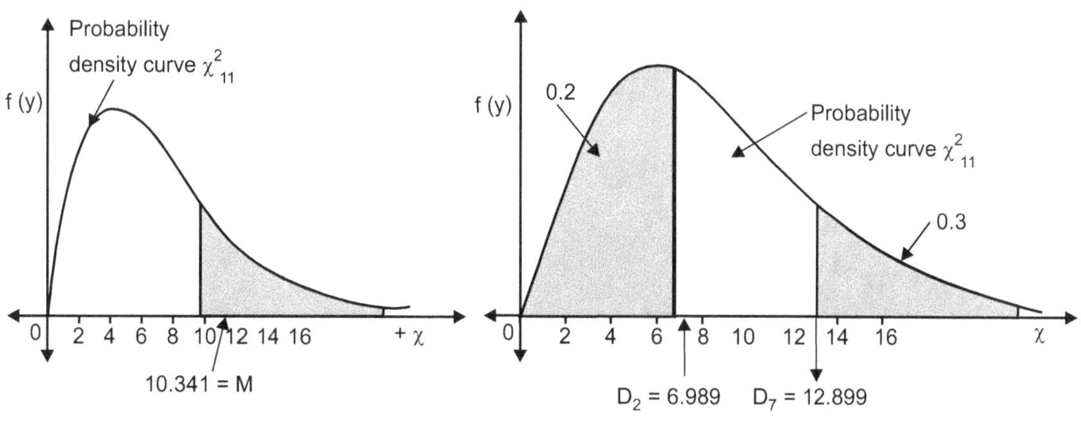

Fig. 1.4　　　　**Fig. 1.5**

(B) USE OF MS-EXCEL IN COMPUTING PROBABILITIES

Suppose we want to find P $[\chi^2_{11} > 8.148]$ using MS-EXCEL.

Step 1 : Click on $\boxed{\text{Insert}}$ on the Excel sheet.

Step 2 : Click on $\boxed{fx}$.

Step 3 : Select function $\boxed{\text{CHIDIST}}$.

Then a box will appear on the screen as follows :

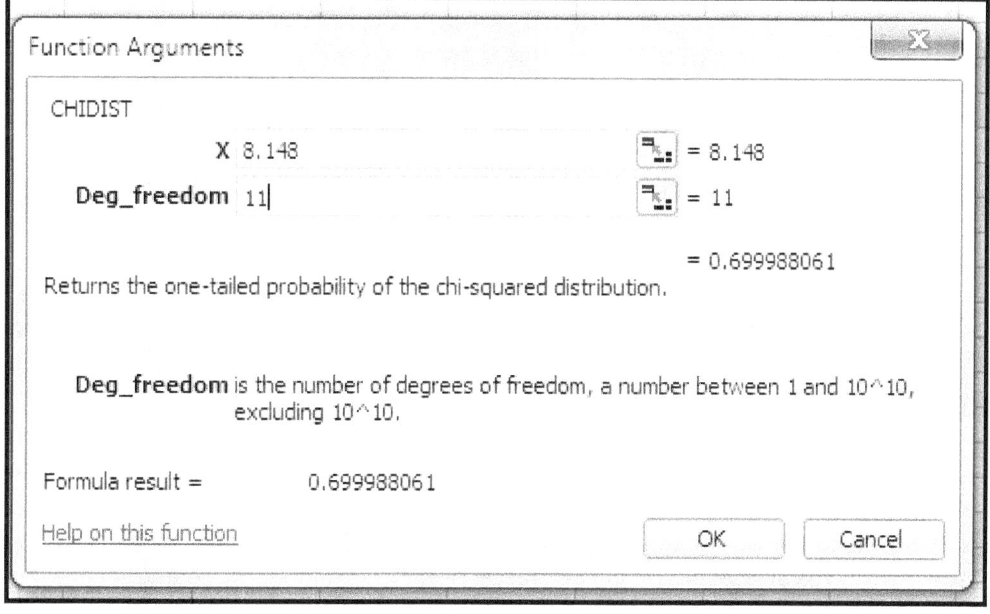

Fig. 1.6 (a)

Enter the value of X = 8.148 and Deg_freedom as 11.

Step 4 : Click on $\boxed{\text{OK}}$.

	A	B	C	D
	A2 ▼	f_x	=CHIDIST(8.148,11)	
1				
2	0.699988			
3				

Fig. 1.6 (b)

The answer 0.699988 appears on the screen.

Note that we always get probability of upper tail.

Note : The above probability can be directly obtained using MS-EXCEL Command $\boxed{= \text{CHIDIST (8.148, 11)}}$. Thus $\boxed{= \text{CHIDIST (x, n)}}$ gives $P\,(\chi_n^2 \geq x)$.

Computation of partition values of χ_n^2 :

(i) To find the median we use the command $\boxed{= \text{CHIINV (0.5, n)}}$ where n is the degrees freedom.

$\boxed{\text{CHIINV (0.5, 11)}}$ gives 10.341. Hence 10.341 is the median of χ_{11}^2.

(ii) To find the first qartile Q_1 we use command $= \boxed{= \text{CHIINV (0.75, n)}}$.

$\boxed{\text{CHIINV (0.75, 11)}}$ gives 7.584143; which is the first quartile of χ_{11}^2.

(iii) To find the second decile D_2 we use command $\boxed{= \text{CHIINV (0.8, n)}}$,

$\boxed{\text{CHIINV (0.8, 11)}}$ gives 6.9887, which is the second decile of χ_{11}^2.

In general $\boxed{= \text{CHIINV (1 − α, n)}}$ gives an ordinate x such that $P\,(\chi_n^2 \geq a) = \alpha$ or $P(\chi^2 \leq a) = 1 - \alpha.$

(C) USE OF R SOFTWARE COMPUTING PROBABILITIES

R software supports following two functions for computing probabilities and ordinates.

(i) pchisq(x,n) : It is used to find $P[X \leq x]$ where X has chi-square distribution with n degrees of freedom.

(a) Suppose X has chi-square distribution with 10 d.f. we want to find $P[X \leq 12.5]$ then we use command.

```
>pchisq(12.5,10)
```

it gives answer 0.7470147

(b) Similarly to find $P[X > 8.148]$ when X has chi-square distribution with 11 d.f. Use command

```
> 1-pchisq(8.148, 11)
```

It gives answer 0.6999881.

(ii) qchisq(p, n) : It is used to find the value of k such that $P[X \le k] = p$ where X has chi-square distribution with n d.f.

    (a) Suppose we want to find k such that $P[X \le k] = 0.78$ when x has chi-square distribution with 12 d.f. we use command

```
> qchisq (0.78,12)
```

    It gives answer 15.40562 which is the value of k.

    (b) Suppose we want to find k such that $P[X \ge k] = 0.58$ when x has chi-square distribution with 15 d.f. Now $P[X \ge k] = 0.58$ gives $1 - P[X \le k] = 0.58$ hence $P[X \le k] = 0.42$ we use command

```
> qchisq (0.42, 15)
```

    It gives answer 13.28882 which is the value of k.

Illustration : To find median, quartiles, deciles we use function qchisq(·). Suppose $Y \to \chi^2_{11}$ then

(i) median is given by command $\boxed{\text{> qchisq (0.5, 11)}}$ = 10.341.

(ii) first quartile Q_1 is given by command $\boxed{\text{+ qchisq (0.25, 11)}}$ = 13.70.

(iii) second decile D_2 is given by command $\boxed{\text{> qchisq (0.2, 11)}}$ = 6.989.

1.7 Limiting Behaviour of χ^2 as $n \to \infty$ [Normal Approximation]

Earlier we have proved central limit theorem in general. It gives the limiting behaviour. We can prove the same taking chi-square random variable as a particular case. We can view χ^2_n as a sum of n i.i.d. χ^2_1, hence stanadrdized χ^2_n will follows N(0, 1) as $n \to \infty$.

Statement : If $Y \to \chi^2_n$ then $Z = \dfrac{Y - E(Y)}{\sqrt{Var(Y)}} = \dfrac{Y - n}{\sqrt{2n}} \to N(0, 1)$ as $n \to \infty$

Proof : It is enough to prove that m.g.f. of Z tends to m.g.f. of N(0, 1) which is $e^{t^2/2}$.

We have,

$$M_Z(t) = M_{\left(\frac{Y-n}{\sqrt{2n}}\right)}(t) = e^{-\frac{nt}{\sqrt{2n}}} M_Y\left(\frac{t}{\sqrt{2n}}\right) = e^{-\left(\sqrt{\frac{n}{2}}\right)t}\left(1 - \frac{2t}{\sqrt{2n}}\right)^{-\frac{n}{2}}$$

$$= e^{-\left(\sqrt{\frac{n}{2}}\right)t}\left(1 - \sqrt{\frac{2}{n}}\,t\right)^{-\frac{n}{2}}$$

Taking logarithms on both sides

$$\ln M_Z(t) = -\left(\sqrt{\frac{n}{2}}\right)t - \frac{n}{2}\ln\left(1 - \sqrt{\frac{2}{n}}\,t\right)$$

$$= -\left(\sqrt{\frac{n}{2}}\right)t + \frac{n}{2}\left\{\sqrt{\frac{2}{n}}\,t + \frac{2}{n}\frac{t^2}{2} + \frac{2}{n}\sqrt{\frac{2}{n}}\frac{t^3}{3} + \dots\right\}$$

$$= -\sqrt{\frac{n}{2}}\,t + \sqrt{\frac{n}{2}}\,t + \frac{t^2}{2} + \sqrt{\frac{2}{n}}\frac{t^3}{3} + \frac{2}{n}\frac{t^4}{4} + \dots\dots$$

$$\therefore \quad \lim_{n \to \infty} \ln M_Z(t) = \frac{t^2}{2}$$

$$\therefore \quad \lim_{n \to \infty} M_Z(t) = e^{t^2/2} \quad \text{which is m.g.f. of N (0, 1). Hence the proof.}$$

Note : There is yet another form of transformation with limiting distribution as N(0, 1). It is known as Fishers approximation.

Fisher's Approximation :

Statement : If $Y \to \chi_n^2$ then $\sqrt{2Y} - \sqrt{2n - 1} \to N(0, 1)$ as $n \to \infty$

$$\text{or } \sqrt{2Y} \to N\left(\sqrt{2n - 1},\ 1\right), \text{ for large } n$$

Proof of this statement is out of scope of this book. However, we use it for solving problems.

$$\textbf{Note :} \quad P[Y \le a] = P\left(\sqrt{2Y} \le \sqrt{2a}\right) = P\left(\sqrt{2Y} - \sqrt{2n - 1} \le \sqrt{2a} - \sqrt{2n - 1}\right)$$

$$= P\left[N(0, 1) \le \sqrt{2a} - \sqrt{2n - 1}\right] = \phi\left(\sqrt{2a} - \sqrt{2n - 1}\right)$$

$$\text{or} \quad P[Y \le a] = P\left[\frac{Y - n}{\sqrt{2n}} \le \frac{a - n}{\sqrt{2n}}\right] \approx \phi\left(\frac{a - n}{\sqrt{2n}}\right)$$

1.8 Results Based on Bivariate Transformations

Some important results involving bivariate transformation are given below :

Result 1. If X and Y are independent chi-square variates with m and n d.f. respectively then,

$$V = \frac{X}{X + Y}$$

$$\text{and} \quad U = X + Y$$

are independently distributed.

Proof : Let

$$U = X + Y$$

$$V = \frac{X}{X + Y}$$

$$\therefore \quad X = UV$$

$$Y = U - UV = U(1 - V)$$

$$\therefore \quad J = \begin{vmatrix} v & u \\ 1 - v & -u \end{vmatrix} = -u; \qquad\qquad (0 < u < \infty,\ 0 < v < 1)$$

$$g(u, v) = f(x, y)\,|J|$$

$$= \frac{\left(\dfrac{1}{2}\right)^{\frac{m}{2} + \frac{n}{2}}}{\Gamma\left(\dfrac{m}{2}\right)\Gamma\left(\dfrac{n}{2}\right)}\, e^{-\left(\frac{x + y}{2}\right)}\, x^{\frac{m}{2} - 1}\, y^{\frac{n}{2} - 1}\, u$$

$$(\because \text{ X and Y are independent})$$

$$= c\, e^{-\frac{u}{2}} (uv)^{\frac{m}{2}-1} [u(1-v)]^{\frac{n}{2}-1} u \qquad \text{where, } C = \frac{\left(\frac{1}{2}\right)^{\frac{m+n}{2}}}{\Gamma\left(\frac{m}{2}\right)\Gamma\left(\frac{n}{2}\right)}$$

$$= c\, e^{\frac{-u}{2}} u^{\frac{m+n}{2}-1} v^{\frac{m}{2}-1} (1-v)^{\frac{n}{2}-1}$$

$$= \left\{ \frac{\left(\frac{1}{2}\right)^{\frac{m+n}{2}}}{\Gamma\left(\frac{m+n}{2}\right)} e^{\frac{-u}{2}} u^{\frac{m+n}{2}-1} \right\} \cdot \left\{ \frac{1}{B\left(\frac{m}{2},\frac{n}{2}\right)} v^{\frac{m}{2}-1} (1-v)^{\frac{n}{2}-1} \right\}$$

$$; 0 < u < \infty,\; 0 < v < 1$$

$$\text{where, } B\left(\frac{m}{2},\frac{n}{2}\right) = \frac{\Gamma\left(\frac{m}{2}\right)\Gamma\left(\frac{n}{2}\right)}{\Gamma\left(\frac{m+n}{2}\right)}$$

We observe that the function of u in the first bracket is the p.d.f. of χ^2 distribution with (m + n) d.f. Thus U and V are independently distributed. In the second bracket we have p.d.f. of beta distribution of first kind.

Note : Similarly, we can prove the result for

$U = X + Y$ and $V = \dfrac{Y}{X + Y}$. In this case $U \to \chi^2_{m+n}$

Result 2 : If X and Y are independent χ^2 variates with m and n d.f. respectively then $X + Y$ and $\dfrac{X}{Y}$ are independently distributed.

Proof : Let $U = X + Y$, $V = \dfrac{X}{Y}$. Then $X = \dfrac{UV}{1 + V}$, $Y = \dfrac{U}{1 + V}$

$$J = \begin{vmatrix} \dfrac{v}{1+v} & \dfrac{u}{(1+v)^2} \\[2mm] \dfrac{1}{1+v} & \dfrac{-u}{(1+v)^2} \end{vmatrix}$$

$$= \frac{-u}{(1+v)^2} \qquad ; \begin{array}{l} 0 < u < \infty \\ 0 < v < \infty \end{array}$$

The joint p.d.f. of U and V is given by,

$$g(u, v) = f(x, y)\,|J| = f_1(x)\, f_2(y)\,|J| \qquad\qquad [\because X \text{ and } Y \text{ are independent}]$$

$$= \left\{ \frac{\left(\frac{1}{2}\right)^{\frac{m}{2}}}{\Gamma\left(\frac{m}{2}\right)} e^{\frac{-x}{2}} x^{\frac{m}{2}-1} \right\} \left\{ \frac{\left(\frac{1}{2}\right)^{\frac{n}{2}}}{\Gamma\left(\frac{n}{2}\right)} e^{\frac{-y}{2}} y^{\frac{n}{2}-1} \right\} \frac{u}{(1+v)^2}$$

$$= \frac{\left(\frac{1}{2}\right)^{\frac{m+n}{2}}}{\Gamma\left(\frac{m}{2}\right)\Gamma\left(\frac{n}{2}\right)} e^{\frac{-(x+y)}{2}} x^{\frac{m}{2}-1} y^{\frac{n}{2}-1} \frac{u}{(1+v)^2}$$

$$= \frac{\left(\frac{1}{2}\right)^{\frac{m+n}{2}}}{B\left(\frac{m}{2},\frac{n}{2}\right) \cdot \Gamma\left(\frac{m+n}{2}\right)} e^{\frac{-u}{2}} \left(\frac{uv}{1+v}\right)^{\frac{m}{2}-1} \left(\frac{u}{1+v}\right)^{\frac{n}{2}-1} \frac{u}{(1+v)^2}$$

$$= \left\{ \frac{\left(\frac{1}{2}\right)^{\frac{m+n}{2}}}{\Gamma\left(\frac{m+n}{2}\right)} \cdot e^{\frac{-u}{2}} u^{\frac{m+n}{2}-1} \right\} \left\{ \frac{1}{B\left(\frac{m}{2},\frac{n}{2}\right)} \frac{v^{\frac{m}{2}-1}}{(1+v)^{\frac{m+n}{2}}} \right\} \begin{array}{l} 0 < u < \infty; \\ 0 < v < \infty \end{array}$$

It can be seen that the function of u is in the first bracket is p.d.f. of χ^2 distribution with (m + n) d.f. In the second bracket we have p.d.f. of beta distribution of second kind.

Thus U and V are independently distributed.

Note : The above result is also true for U = X + Y and V = $\frac{Y}{X}$.

In this case U $\rightarrow \chi^2_{m+n}$.

Result 3 : If X is a continuous r.v. with probability density function

$$f(x) = \frac{1}{B(m, n)} x^{m-1} (1-x)^{n-1} \qquad ; \qquad 0 \le x \le 1 \text{ with } m > 0 \text{ and } n > 0;$$

$$= 0 \qquad ; \qquad \text{otherwise}$$

then for large m and comparatively small n, $- 2m \log_e X$ follows approximately χ^2 distribution with 2n d.f.

Proof :   Let, $U = - 2m \log_e X$

$$= 2 mt \text{ where, } t = - \log_e X$$

$$\therefore X = e^{-t} \text{ and } t = \frac{U}{2m}$$

Hence the p.d.f. of U is

$$h(u) = g(t) \left| \frac{dt}{du} \right|$$

$$= g(t) \cdot \frac{1}{2m} ; m > 0$$

Here the p.d.f. of g(t) is given by,

$$g(t) = f(x) \left| \frac{dx}{dt} \right|$$

$$= \frac{1}{B(m, n)} \, x^{m-1} (1-x)^{n-1} e^{-t}$$

$$= \frac{\Gamma(m+n)}{\Gamma(m) \, \Gamma(n)} \, e^{-mt} (1 - e^{-t})^{n-1}$$

As $m \to \infty$ and for constant n, $\dfrac{\Gamma(m+n)}{\Gamma(m)} \approx m^n$

$$\therefore \qquad g(t) = \frac{m^n}{\Gamma(n)} \, e^{-mt} \left\{ 1 - \left[1 - t + \frac{t^2}{2!} - \frac{t^3}{3!} + \cdots \right] \right\}^{n-1}$$

$$= \frac{m^n}{\Gamma(n)} \, e^{-mt} \, t^{n-1} \left[1 - t + \frac{t^2}{2!} - \frac{t^3}{3!} + \cdots \right]^{n-1}$$

Since n is comparatively small, using the first approximation we get,

$$g(t) = \frac{m^n}{\Gamma(n)} \, e^{-mt} \, t^{n-1} ; t \geq 0$$

$\therefore$ The p.d.f. of U is given by,

$$h(u) = \frac{m^n}{\Gamma(n)} \, e^{-mt} \, t^{n-1} \cdot \frac{1}{2m} = \frac{m^n}{\Gamma(n)} \, e^{\frac{-u}{2}} \left(\frac{u}{2m} \right)^{n-1} t^{n-1} \cdot \frac{1}{2m}$$

$$h(u) = \frac{1}{2^n \, \Gamma(n)} \, e^{\frac{-u}{2}} \, u^{n-1}$$

$$h(u) = \frac{\left(\frac{1}{2} \right)^{\frac{2n}{2}}}{\Gamma \left(\frac{2n}{2} \right)} \cdot e^{\frac{-u}{2}} \, u^{\left(\frac{2n}{2} \right) - 1} \qquad ; u \geq 0$$

Thus h(u) is the p.d.f. of χ^2 distribution with (2n) d.f.

Solved Examples

Example 1.1 : If $Y \to \chi^2_{16}$, find (i) P [Y $\leq$ 20.465] (ii) P [11.152 $\leq$ Y $\leq$ 26.296] (iii) median of Y.

Solution :

(i)   P[Y $\leq$ 20.465] $= 1 - P[Y > 20.465]$

$$= 1 - P \left[\chi^2_{16} > 20.465 \right] = 1 - 0.2$$

$$= 0.8$$

(ii) $P[11.152 \leq Y \leq 26.296] = P[Y \geq 11.152] - P[Y \geq 26.296] = 0.8 - 0.05 = 0.75$

(iii)   Suppose median is M. Then $P[Y \geq M] = 0.8$

$$\therefore \qquad P\left[\chi^2_{16} \geq M\right] = 0.5$$

$$\therefore \qquad\qquad M = 15.338$$

Example 1.2 : If $X_1, X_2, \ldots\ldots X_{113}$ are i.i.d. standard normal variates calculate :

$$P\left[\sum_{i=1}^{113} X_i^2 \leq 128\right]$$

Solution : By additive property of χ^2 variates

$$Y = \sum_{i=1}^{113} X_i^2 \rightarrow \chi^2_{113}$$

$$\therefore \qquad P\left[\sum_{i=1}^{113} X_i^2 \leq 128\right] = P\left[\chi^3_{113} \leq 128\right]$$

Since table values for χ^2_{113} are not available, we use normal approximation.

$$= P\left[\sqrt{2Y} \leq \sqrt{256}\right] \text{ By Fisher's approximation}$$

$$= P\left[\sqrt{2Y} - \sqrt{2n-1} \leq \sqrt{256} - \sqrt{226-1}\right]$$

$$= P[N(0, 1) \leq 16 - 15]$$

$$= P[N(0, 1) \leq 1]$$

$$= 1 - 0.15866 = 0.84134$$

Example 1.3 : Let $X_1, X_2 \ldots X_{20}$ be a random sample from normal population with mean $\mu = 5$ and variance $\sigma^2 = 16$

Determine $P\left[\sum_{i=1}^{20} (X_i - 5)^2 \geq 364.4\right]$

Solution :   Let, $Y = \sum_{i=1}^{20} \left(\dfrac{X_i - 5}{4}\right)^2 = \dfrac{\sum_{i=1}^{20} (X_i - 5)^2}{16}$

$\because X_i \rightarrow N(\mu = 5, \sigma^2 = 4^2), i = 1 \ldots 20$

$$Y \rightarrow \chi^2_{20}$$

$$\therefore \quad P\left[\sum_{i=1}^{20} (X_i - 5)^2 \geq 364.4\right] = P\left[\dfrac{\sum_{i=1}^{20} (X_i - 5)^2}{16} \geq \dfrac{364.4}{16}\right] = P\left[\chi^2_{20} \geq 22.775\right]$$

$$= 0.30$$

Example 1.4 : Let $X_1, X_2, \dots X_i \dots X_{10}$ be independent normal variates such that $E(X_i) = 0$ and $Var(X_i) = i^2$.

Find $P\left[X_1^2 + \dfrac{X_2^2}{2^2} + \dfrac{X_3^2}{3^2} + \dots + \dfrac{X_{10}^2}{10^2} \geq 7.267 \right]$

Solution : $\because X_i \to N(0, i^2), \quad \dfrac{X_i^2}{i^2} \to \chi_1^2$

$\therefore \qquad \displaystyle\sum_{i=1}^{10} \dfrac{X_i^2}{i^2} \to \chi_{10}^2$

$\therefore \quad P\left[X_1^2 + \dfrac{X_2^2}{2^2} + \dfrac{X_3^2}{3^2} + \dots + \dfrac{X_{10}^2}{10^2} \geq 7.267 \right] = P\left[\displaystyle\sum_{i=1}^{10} \dfrac{X_i^2}{i^2} \geq 7.267 \right] = P\left[\chi_{10}^2 \geq 7.267 \right]$

$$= 0.7$$

Example 1.5 : If $X_1 \to N(4, 1)$, $X_2 \to N(4, 1)$ and X_1, X_2 are independent, find

$\qquad P[0.296 < (X_1 - X_2)^2 < 0.910]$ **(P.U. April 2000)**

Solution : $X_1 - X_2 \to N(0, 2)$

$\therefore \qquad\qquad Y = \dfrac{(X_1 - X_2)^2}{2} \to \chi_1^2$

$\therefore \quad P[0.296 < (X_1 - X_2)^2 < 0.910] = P\left[\dfrac{0.296}{2} < \dfrac{(X_1 - X_2)^2}{2} < \dfrac{0.910}{2} \right]$

$$= P\left[0.148 < \chi_1^2 < 0.455\right]$$

$$= P\left[\chi_1^2 \geq 0.148\right] - P\left[\chi_1^2 \geq 0.455\right]$$

$$= 0.7 - 0.5$$

$$= 0.2$$

Example 1.6 : If $X \to \chi_2^2$, $Y \to \chi_2^2$ and X, Y are independent then find

(i) $P\left(X + Y \leq 1.064, \dfrac{X}{Y} > 1 \right)$

(ii) $\left[X + Y \leq 2.195, \dfrac{X}{X + Y} < \dfrac{1}{2} \right]$

Solution : (i) Let, $U = X + Y$ $Y = \dfrac{X}{Y}$ By result (2), U and V are independent. Also $U \to \chi^2_4$ and V has p.d.f. given by,

$$f(v) \;=\; \frac{1}{B(1,1)} \cdot \frac{1}{(1+v)^2} \qquad ;\; 0 < v < \times$$

$$\therefore P\left(X + Y \le 1.064, \frac{X}{Y} > 1\right) = P(U \le 1.064, V > 1)$$

$$=\; P(U \le 1.064) \cdot P[V > 1] \qquad\qquad \because \text{U and V are independent.}$$

$$=\; P\left[\chi^2_4 \le 1.064\right] \cdot \frac{1}{B(1,1)} \int_{1}^{\infty} \frac{1}{(1+v)^2}\, dv$$

$$=\; (0.1)\left[-(1+v)^{-1}\right]_{1}^{\infty}$$

$$=\; \frac{0.1}{2}$$

$$=\; 0.05$$

(ii) If $U = X + Y$ and $V = \dfrac{X}{X+Y}$ then by result (i), U and V are independent. Also $U \to \chi^2_4$ and V has p.d.f. given by,

$$f(v) \;=\; \frac{1}{B(1,1)}\, v^{1-1}(1-v)^{1-1} = \frac{1}{B(1,1)} = 1 \;;\; 0 < v < 1$$

$$\therefore \qquad\qquad v \;\to\; U(0,1)$$

$$\therefore\; P\left[X + Y \le 2.195, \frac{X}{X+Y} < \frac{1}{2}\right] = P\left[U \le 2.195, V < \frac{1}{2}\right]$$

$$=\; P[U \le 2.195]\, P\left[V < \frac{1}{2}\right] \qquad (\because \text{U and V are independent})$$

$$=\; P[U \le 2.195] \int_{0}^{1/2} dv = (0.3) \cdot \left(\frac{1}{2}\right)$$

$$=\; 0.15$$

Example 1.7 : If $X_1 \to N(0,1)$, $X_2 \to N(0,1)$ and X_1, X_2 are independent then find

$$P\left[(X_1 + X_2)^2 \le 3.284, (X_1 - X_2)^2 \ge 2.148\right]$$

Solution : If X_1, X_2 are two independent standard normal variates then

$\dfrac{(X_1 + X_2)^2}{2}$ and $\dfrac{(X_1 - X_2)^2}{2}$ are independent χ^2 variates with 1 d.f.

$$\therefore \quad P\left[(X_1 + X_2)^2 \le 3.284,\ (X_1 - X_2)^2 \ge 2.148\right]$$

$$= P\left[\frac{(X_1 + X_2)^2}{2} \le 1.642,\ \frac{(X_1 - X_2)^2}{2} \ge 1.074\right]$$

$$= P\left[\chi_1^2 \le 1.642\right] \cdot P\left[\chi_1^2 \ge 1.074\right] \qquad \because \left[\frac{(X_1 + X_2)^2}{2} \text{ and } \frac{(X_1 - X_2)^2}{2} \text{ are independent}\right]$$

$$= (0.8)\,(0.3)$$

$$= (0.24)$$

Points to Remember

1. The p.d.f. chi-square distribution with n d.f. is

$$f(y) = \frac{\left(\frac{1}{2}\right)^{n/2}}{\Gamma\left(\frac{n}{2}\right)}\ e^{-y/2}\ y^{\frac{n}{2} - 1}\ ;\ y \ge 0.$$

2. If $y \to \chi_n^2$ then

    (a) First 4 raw moments are given by

$$\mu_1' = \text{mean} = n,\quad \mu_2' = n\,(n + 1),\quad \mu_3' = n\,(n + 2)\,(n + 4),\quad \mu_4' = n\,(n + 2)\,(n + 4)\,(n + 6).$$

    (b) First 4 central moments are given by

$$\mu_1 = 0,\quad \mu_2 = 2n,\quad \mu_3 = 8n,\quad \mu_4 = 48n + 12n^2.$$

    (c) $\beta_1 = \dfrac{8}{n},\quad \gamma_1 = \sqrt{\dfrac{8}{n}},\quad \beta_2 = 3 + \dfrac{12}{n},\quad \gamma_2 = \dfrac{12}{n}.$

    (d) The rth raw moments is $\mu_r' = n\,(n + 2) \ldots (n + 2r - 2)\ ;\ r = 1, 2 \ldots$

    (e) The moment generating function (m.g.f.) is given by

$$M_X(t)\ =\ (1 - 2t)^{-n/2}\ ;\ t < \frac{1}{2}$$

    (f) The cumulant generating function (c.g.f.) is

$$K_X(t)\ =\ -\frac{n}{2}\,ln\,(1 - 2t) \text{ and } r^{th} \text{ cumulant}$$

$$K_r\ =\ (r - 1)\,!\,n\,2^{r-1}\,;\ r = 1, 2 \ldots$$

    (g) The recurrence relation among the central moments is

$$\mu_{r+1}\ =\ 2r\,[n\,\mu_{r-1} + \mu_r]\ ;\ r = 1, 2, 3 \ldots$$

(h) The mode of the distribution is $n - 2$ for $n > 2$.

(i) $\dfrac{Y - E(Y)}{\sqrt{\text{Var}(Y)}} = \dfrac{Y - n}{\sqrt{2n}} \to N(0, 1)$ as $n \to \infty$ [normal approximation]

(j) $\sqrt{2Y} - \sqrt{2n - 1} \to N(0, 1)$ for large n [Fisher's approximation].

3. **Additive property :** If Y_1 and Y_2 are independent χ^2 variates with n_1 and n_2 d.f. respectively then $Y_1 + Y_2$ has also χ^2 distribution with $(n_1 + n_2)$ degrees of freedom. The generalization of this property also holds good.

4. **Important Results :**

(i) If X and Y are independent chi-square variates with m and n d.f. respectively then $V = \dfrac{X}{X + Y}$ and $U = X + Y$ are independently distributed. Similarly $U = X + Y$ and $V = \dfrac{Y}{X + Y}$ are independently distributed. In this case $U \to \chi^2_{m + n}$.

(ii) If X and Y are independent variates with m and n d.f. respectively then $U = X + Y$ and $V = \dfrac{X}{Y}$ are independently distributed. Similarly $U = X + Y$ and $V = \dfrac{Y}{X}$ are also independently distributed. Here $U = \chi^2_{m + n}$.

(iii) If X is a continuous random variable with probability density function

$$f(x) = \dfrac{1}{B(m, n)} \; X^{m - 1} (1 - X)^{n - 1} \quad ; \quad 0 \le x \le 1 \text{ with } m > 0, n > 0$$

$$= 0 \qquad\qquad\qquad ; \quad \text{otherwise}$$

then for large m and comparatively small n, $-2m \log_e X$ follows approximately χ^2 distribution with 2n d.f.

Exercise 1 (A)

(1) (a) Let $X_1, X_2, ..., X_n$ be i.i.d. $N(0, 1)$ variates. Obtain the moment generating function of $Y = \sum\limits_{i=1}^{n} X_i^2$. Identify the distribution of Y.

(b) Define a chi-square with n d.f. and derive its p.d.f.

(P.U. May 2001, Oct. 2006, 2012)

(2) (a) Define χ^2 variate with n degrees of freedom. Find its mean and variance.

(P.U. May 2003)

(b) Obtain the rth raw moment of χ^2 distribution with n d.f. Hence, find mean and variance.　　**(P.U. April 2004, Nov. 2004)**

(3) Obtain the m.g.f. and hence find γ_1 and γ_2 for a chi-square distribution.

(P.U. April 2004)

(4) State and prove additive property for χ^2 variates.   **(P.U. April 2013)**

(5) Obtain Karl Pearson's coefficient of skewness for a chi-square distribution.

(P.U. April 98)

(6) Show that the mode of the chi-square distribution with n d.f. is n – 2.

Also find P [X > 2.343] if mode of the distribution is 3.

(P.U. Oct. 2011) (P.U. April 99)

(7) If X and Y are independent chi-square variates with m and n d.f. respectively, show

that $\dfrac{X}{X + Y}$ and X + Y are independently distributed.

(8) Show that if X and Y are independent χ^2 variate with m and n d.f. respectively then

X + Y and $\dfrac{X}{Y}$ are independently distributed.   **(P.U. April 2000, P.U. Oct. 98)**

(9) If X is a continuous r.v. with probability density function

$f(x) = \dfrac{1}{B\,(m,\,n)}\ x^{m-1}\,(1-x)^{n-1}\,;\ 0 \le x \le 1$ with m > 0, n > 0

$ = 0 ;$ otherwise

then show that for large m and comparatively small n, $-\,2m\,\log_e X$ follows approximately χ^2 distribution with 2n d.f.

(10) Show that for a χ^2 variate with n d.f.

$\mu_{r+1} = 2r\,[n\,\mu_{r-1} + \mu_r]\,;\ r = 1, 2, 3 \ldots$ and hence obtain first four central moments

of the distribution.   **(P.U. 2005)**

(11) (a) Derive the recurrence relation between raw moments of chi-square distribution.

(b) Let X be a chi-square variate with n d.f. Find the m.g.f. of $y = \dfrac{X - n}{\sqrt{2n}}$. Hence,

find its limiting values as $n \to \infty$ and interpret the result.

(P.U. May 2011, May 2003)

(12) Let X_1 and X_2 be two independent N(0, 1) variates. Show that $Y_1 = \dfrac{(X_1 + X_2)^2}{2}$ and

$Y_2 = \dfrac{(X_1 - X_2)^2}{2}$ are independent chi-square variates with 1 d.f. each.

(13) Obtain cumulant generating function (c.g.f) of chi-square distribution. Hence $\beta_1,\ \gamma_1,$

β_2 and γ_2.   **(P.U. October 2005)**

(14) State limiting behaviour of χ_n^2 as $n \to \infty$ according to (i) Normal approximation.

(ii) Fisher's approximation.

(15) Let $X_1, X_2 \ldots X_6$ be i.i.d. N(0, 1) variates. State giving reasons the distribution of

$$U = \dfrac{X_1^2 + X_2^2 + X_3^2}{X_4^2 + X_5^2 + X_6^2}$$

(16) If X and Y are two independent chi-square variates with m and n d.f. respectively. State the distribution of (i) X + Y. (ii) $\sqrt{2X} - \sqrt{2m-1}$ for large m.

(17) Let $X_1, X_2 \ldots X_m, X_{m+1}, X_{m+2} \ldots X_{m+n}$ be i.i.d. standard normal variates. If

$$U = \sum_{i=1}^{m} X_i^2 \text{ and } V = \sum_{j=1}^{n} X_{m+j}^2 \text{, obtain the distribution of } W = \frac{nU}{mV}.$$

(18) Let X have a standard normal distribution. Then derive the distribution of $Y = X^2$.

(19) Let Y_i (i = 1, 2 … n) be n independent chi-square variates with i^2 k d.f. state the

distribution of $\displaystyle\sum_{i=1}^{n-1} (Y_i - i\,Y_n) + \frac{n(n-1)}{2} Y_n$

(20) Let the p.d.f. of r.v. X be given by,

$$f(x) = (2\pi)^{-1/2}\, e^{\frac{-x^2}{2}} \;\; ; -\infty < x < \infty$$

Show that $Y = X^2$ follows chi-square distribution with 1 degree of freedom.

(21) Discuss the nature of probability curves for chi-square distribution for n = 1, n = 2 and n > 2, where n is degrees of freedom. Write the type of skewness for the curve when n < 2.

(22) Establish the recurrence relation among the central moments of χ^2 distribution with n d.f.

(23) Let X and Y be two independent chi-square variates with 10 and 20 d.f. respectively. Find median of the probability distribution of X + Y.

(24) The probability density of continuous random variable X is given by

$$f(x) \;\; = \;\; \frac{1}{\sqrt{2\pi}}\, e^{x^2/2} \;\; ; -\infty < x < \infty.$$

Find the distribution of X^2. If $X_1, X_2, \ldots, X_n$ are n independent standard normal variates then state the distribution of $Y = \sum_{i=1}^{n} X_i^2$. 　　　**[P.U. 2011]**

Exercise 1 (B)

(1) Let $X_1, X_2 \ldots X_{60}$ be i.i.d. N(0, 1) variates. Find $P\left[\displaystyle\sum_{t=1}^{60} X_t^2 < 65.227\right]$

(2) If X_1 and X_2 are two independent N (0, 1) and N $\left(0, \dfrac{1}{2}\right)$ variates. State the probability distribution of $X_1^2 + 2X_2^2$.

(3) Let X_i (i = 1, 2, …, 145) be i.i.d. N (0, 2) variates Find $P\left(X_1^2 + X_2^2 + \ldots + X_{145}^2 \leq 162\right)$

(Nov. 2004)

(4) Identify the distribution of a r.v. X if its m.g.f.

$$M_X(t) = (1 - 2t)^{-8} \text{ where } t < \frac{1}{2}.$$

(5) If $M_X(t) = \left(1 - \dfrac{t}{1/2}\right)^{-20}$ where $t < \dfrac{1}{2}$ is the m.g.f. of a r.v. X. find median of X.

(6) Show that for a χ^2 distribution with 2 d.f. $P_0 = P[\chi^2 \geq v] = e^{-u/2}$. If $P_0 = 0.05$ find v.

(P.U. April 99)

(7) (a) Let $X_1, X_2, ..., X_{10}$ be i.i.d. N(5, 10) variates. Calculate $P\left[\sum\limits_{i=1}^{10} (X_i - 5)^2 \geq 72.67\right]$.

(P.U. April 98)

(b) Let X_i, $i = 1, 2 ... 8$ be independent and identically distributed

N ($\mu = 20$, $\sigma^2 = 20$) variates. Calculate $P\left[\sum\limits_{i=1}^{8} (X_i - 20)^2 \geq 190.48\right]$.

(P.U. May 2003)

(8) Let $X_1, X_2 ... X_8$ be a random sample from N(4, 25) distribution. Find mean and median of Y where,

$$Y = \sum_{i=1}^{8} (X_i - 4)^2$$

(9) If X_1 and X_2 are two independent normal variates with mean 4 and S.D. 1 each find
P [0.296 < $(X_1 - X_2)^2$ < 0.910] **(P.U. Oct. 98)**

(10) Let $X_1, X_2, ..., X_r, ..., X_{12}$ be independent normal variates such that E $(X_r) = 0$ and

Var $(X_r) = r^2$. Find $P\left[X_1^2 + \dfrac{X_2^2}{2^2} + ... + \dfrac{X_r^2}{r^2} + ... + \dfrac{X_{12}^2}{12^2} \leq 11.340\right]$

(11) Let $X_1 \rightarrow N(5.5, 1)$, $X_2 \rightarrow N(5.5, 1)$ and X_1, Y_2 are independent find
P [2.148 < $(X_1 - X_2)^2$ < 5.412]

(12) If X_1, X_2 are independent standard normal variates then find,

P $[(X_1 + X_2)^2 \leq 0.296, (X_1 - X_2)^2 \geq 5.412]$

Objective Questions

Exercise 1 (C)

Choose correct alternative out of (a) to (d) for the following :

(1) If Y follows chi-square distribution with variance 6 then mean of the distribution is
    (a) 4   (b) 12   (c) 2   (d) 3

(2) If X has χ^2 distribution with $\mu'_1 = 6$ then the third row moment μ'_3 is
    (a) 480   (b) 48   (c) 460   (d) 336

(3) If a random variable has χ^2 distribution with variance 4 then its m.g.f. is given by
    (a) $(1 - 2t)^{-1}$ (b) $(1 - 2t)^{-2}$ (c) $(1 - t)^{-1}$ (d) $(1 - t)^{-2}$

(4) If m.g.f. of a continuous random variable Y is given by $M_Y(t) = (1 - 2t)^{-3}$.
Hence, Var (y) is

    (a) 6   (b) 4   (c) 12   (d) 3

(5) The m.g.f. of a continuous random variable X is $M_X(t) = (1 - 2t)^{-4}$. Hence, mode of
the distribution of X is

    (a) 8   (b) 4   (c) 7   (d) 6

(6) If a random variable X has χ^2 distribution with third cumulant 24 then mode of the
distribution is

    (a) 8   (b) 1   (c) 3   (d) 6

(7) If a random variable has χ^2 distribution with 16 d.f. then the coefficient of skewness
β_1 is

    (a) $\dfrac{1}{4}$ (b) $\dfrac{3}{4}$ (c) 1   (d) $\dfrac{1}{2}$

(8) Suppose Y has χ^2 distribution and coefficient of Kurtosis β_2 of its distribution is 6
then E(Y) is

    (a) 4   (b) 8   (c) 6   (d) 5

(9) Let Y_1, Y_2, Y_3 be independent chi-square variates with 8, 4 and 9 d.f. respectively.
$T = Y_1 + Y_2 + Y_3$. Then mode of T is

    (a) 21   (b) 20   (c) 22   (d) 19

(10) If X has χ^2 distribution with mode 12 then it's m.g.f. $M_X(t)$ is

    (a) $(1 - 2t)^{-6}$ (b) $(1 - 2t)^{-12}$ (c) $(1 - 2t)^{-7}$ (d) $(1 - 2t)^{-10}$

(11) If a continuous random variable has χ^2 distribution with mode 13 then its variance
is

    (a) 15   (b) 30   (c) 26   (d) 22

(12) If $Y \rightarrow \chi^2_{61}$ then by Fisher's approximation which one of the following will follow
N (0, 1) distribution ?

    (a) $\sqrt{2Y} - \sqrt{61}$ (b) $\sqrt{2Y} + 11$ (c) $\sqrt{2Y} - \sqrt{122}$ (d) $\sqrt{2Y} - 11$

(13) If $X_1 \rightarrow N(0, 1)$, $X_2 \rightarrow N(0, 1)$ and X_1 and X_2 are independent then distribution of
$\dfrac{(X_1 + X_2)^2}{2}$ is

    (a) χ^2_2 (b) t_2 (c) t_1 (d) χ^2_1

(14) If $X_1 \rightarrow N(4, 4)$, $X_2 \rightarrow N(4, 5)$ are independent variates then the distribution of
$\dfrac{(X_1 - X_2)^2}{9}$ is

    (a) χ^2_1 (b) χ^2_2 (c) χ^2_3 (d) N (0, 1)

(15) The variance of chi-square distribution with n degrees of freedom is

    (a) $\dfrac{n}{n-2}$ (b) n   (c) $\dfrac{n}{2}$ (d) 2 n

(16) If X_1, X_2 are independent random variables following N (0, 1) and N $\left(0, \dfrac{1}{2}\right)$

distribution respectively then probability distribution of $X_1^2 + 2X_2^2$ is

(a) N (3, 5)
(b) F (1, 2)

(c) χ_2^2
(d) Cannot be determined.

(17) If X follows chi-square distribution with variance 22 then median of the distribution is

(a) 11
(b) 10.521
(c) 10
(d) 10.341

(18) A chi-square random variable has variance 12 then its mode is ......

(a) 6
(b) 24

(c) 4
(d) does not exist

Exercise 1 (D)

State True or False :

(19) The p.d.f. of χ^2 distribution with n (> 1) d.f. is same as that of G $\left(\dfrac{1}{2}, \dfrac{1}{2}\right)$ distribution.

(20) The p.d.f. of χ^2 distribution with n d.f. is same as that of G $\left(\dfrac{1}{2}, \dfrac{n}{2}\right)$ distribution.

(21) The p.d.f. of χ^2 distribution with n d.f. is same as that of G $\left(\dfrac{n}{2}, \dfrac{1}{2}\right)$.

(22) The rth raw moment of χ^2 distribution with n d.f. is given by $\dfrac{\Gamma\left(\dfrac{n}{2} + r\right)}{\Gamma\left(\dfrac{n}{2}\right)} \cdot 2^n$.

(23) For χ^2 distribution with n d.f. the rth raw moment is given by $\dfrac{\Gamma\left(\dfrac{n}{2} + r\right)}{\Gamma\left(\dfrac{n}{2}\right)} 2^r$.

(24) χ^2 distribution is symmetric.

(25) The moment generating function of chi-square distribution with 10 degrees of freedom is $(1 - 2t)^{-5}$.

(26) The recurrence relation among the central moments of χ^2 distribution with n d.f. is $\mu_{r+1} = 2n\,[r\,\mu_{r-1} + \mu_r]$ for r = 1, 2, 3 ...

(27) The recurrence relation among the central moments of χ^2 distribution with n d.f. is $\mu_{r+1} = 2r\,[n\,\mu_{r-1} + \mu_r]$ for r = 1, 2, 3 ...

(28) Chi-square probability distribution satisfies additive property.

(29) If $X_1, X_2, ..., X_i, ..., X_{10}$ are independent normal variates such that E (X_i) = 0 and

$$\text{Var } (X_i) = i^2, i = 1, 2, ..., 10 \text{ then P}\left[X_1^2 + \frac{X_2^2}{2^2} + \frac{X_3^2}{3^2} + ... + \frac{X_{10}^2}{10^2} \geq 7.267 \right] = 0.7.$$

(30) Let $X_1, X_2, ..., X_i, ..., X_{10}$ be a random sample with population mean μ and variance

σ^2. Then $\dfrac{\sum\limits_{i=1}^{10} (X_i - \bar{X})^2}{\sigma^2}$ has χ^2 distribution with 10 d.f. (where $\bar{X}$ is sample mean).

(31) The p.d.f. of G $\left(\dfrac{1}{2}, \dfrac{1}{2}\right)$ is same as that of χ^2 distribution with 1 d.f.

(32) If $X_1, X_2, ..., X_{10}$ are independent identically distributed N (0, 1) then the probability

distribution of $\sum\limits_{i=1}^{5} X_i^2$ has chi-square distribution with 4 d.f.

Answers of Exercise 1 (B)

(1) 0.7

(2) χ_2^2

(3) P [N(0, 1) ≤ 1] = 0.84134

(4) χ^2 distribution with 16 d.f.

(5) Median = 39.335

(6) ν = 3.012

(7) 0.7

(8) E(Y) = $\dfrac{8}{25}$

(9) 0.2

(10) 0.5

(11) 0.2

(12) 0.003

Answers of Exercise 1 (C)

(1) d

(2) a

(3) a

(4) c

(5) d

(6) b

(7) d

(8) a

(9) d

(10) c

(11) b

(12) d

(13) c

(14) a

(15) d

(16) c

(17) d

(18) c

Answers of Exercise 1 (D)

(19) False

(20) True

(21) False

(22) False

(23) True

(24) False

(25) True

(26) False

(27) True

(28) True

(29) True

(30) False

(31) True

(32) False

Chapter **2**...

Student's t-Distribution

Willam S. Gossset

Born in Canterbury, England to Agnes Sealy Vidal and Colonel Frederic Gosset, Gosset attended Winchester College before reading chemistry and mathematics at New College, Oxford. Upon graduating in 1899, he joined the brewery of Arthur Guinness & Son in Dublin, Ireland.

As an employee of Guinness, a progressive agro-chemical business, Gosset applied his statistical knowledge – both in the brewery and on the farm – to the selection of the best yielding varieties of barley. Gosset acquired that knowledge by study, by trial and error, and by spending two terms in 1906–1907 in the biometrical laboratory of Karl Pearson. Gosset and Pearson had a good relationship. Pearson helped Gosset with the mathematics of his papers, including the 1908 paper, but had little appreciation of their importance. The papers addressed the brewer's concern with small samples; biometricians like Pearson, on the other hand, typically had hundreds of observations and saw no urgency in developing small-sample methods.

Another researcher at Guinness had previously published a paper containing trade secrets of the Guinness brewery. To prevent further disclosure of confidential information, Guinness prohibited its employees from publishing any papers regardless of the contained information. However, after pleading with the brewery and explaining that his mathematical and philosophical conclusions were of no possible practical use to competing brewers, he was allowed to publish them, but under a pseudonym ("Student"), to avoid difficulties with the rest of the staff.[1] Thus his most noteworthy achievement is now called Student's, rather than Gosset's, t-distribution.

Contents ...

Key Words :

Students t-distribution, limiting distribution of t_n.

Objectives :

(1) To understand how t-distribution is derived from chi-square and standard normal variates.

(2) To study the properties of t-distribution.

(3) To note relationships between t distribution and Cauchy distribution.

2.0 Introduction

    t-distribution is important due to its popularly used applications in statistical inference such as tests of hypothesis regarding mean, correlation coefficient, regression coefficients, determination of confidence interval for mean etc. Some of the applications will be discussed later in this book. William S. Gosset has published the paper on t-distribution in 1908 under the pseudonym 'student' hence it is called as Student's t-distribution.

2.1 P.D.F. of t-Distribution

    Definition : Suppose U and V are independent random variables. If U follows N (0, 1) and V follows chi-square distribution with n degrees of freedom then $t = U/\sqrt{V/n}$ is said to follow t-distribution with parameter n, which is called as degrees of freedom.

$$\left[\text{i.e. } t = \frac{N(0, 1)}{\sqrt{\chi_n^2/n}} \right]. \text{ We symbolically write it as } t_n.$$

Derivation of p.d.f. of t_n :

    Since t is a transformation of U and V we find its p.d.f. as follows :

(i)   Let $t = \dfrac{U}{\sqrt{V/n}}$, $Z = V$

Clearly, $-\infty < t < \infty$ and $0 < Z < \infty$

(ii)   We get, $U = t\sqrt{\dfrac{Z}{n}}$ and $V = Z$

Hence, the Jacobian of transformation is

$$J = \frac{\partial(U, V)}{\partial(t, Z)} = \begin{vmatrix} \dfrac{\partial U}{\partial t} & \dfrac{\partial U}{\partial Z} \\[2ex] \dfrac{\partial V}{\partial t} & \dfrac{\partial V}{\partial Z} \end{vmatrix} = \begin{vmatrix} \sqrt{z/n} & \dfrac{t}{\sqrt{n}} \cdot \dfrac{1}{2\sqrt{z}} \\[2ex] 0 & 1 \end{vmatrix} = \sqrt{\dfrac{z}{n}}$$

(iii)   Joint p.d.f. of (t, Z) = Joint p.d.f. of (U, V) . |J|

$$g(t, Z) = f(u, v) \cdot |J|$$

$$= f_1(u) \cdot f_2(v) |J| \quad (\because U \text{ and } V \text{ are independent})$$

$$= \frac{1}{\sqrt{2\pi}} e^{-u^2/2} \cdot \frac{1}{2^{n/2}\left|\dfrac{n}{2}\right.} e^{-\frac{v}{2}} v^{n/2 - 1} \cdot |J|$$

$$= \frac{1}{\sqrt{2\pi}\ 2^{n/2}\ \left\lfloor \dfrac{n}{2} \right.}\ .\ e^{-\left(\frac{u^2}{2} + \frac{v}{2}\right)}\ .\ v^{\frac{n}{2}-1}\ .\ |J|$$

$$= \frac{1}{\sqrt{2\pi}\ 2^{n/2}\ \left\lfloor \dfrac{n}{2} \right.}\ e^{-\frac{1}{2}\left(Z + \frac{t^2 z}{n}\right)}\ Z^{\frac{n}{2}-1}\ .\ \sqrt{\frac{Z}{n}}$$

$$= \frac{1}{\sqrt{2\pi n}\ 2^{n/2}\ \left\lfloor \dfrac{n}{2} \right.}\ e^{-\frac{z}{2}\left(1 + \frac{t^2}{n}\right)}\ .\ Z^{\frac{n+1}{2}-1}\ ; z > 0,\ -\infty < t < \infty$$

(iv)   To get p.d.f. of t we integrate g (t, Z) w.r.t. z,

$$p\,(t) = \int_{0}^{\infty} g\,(t,\,z)\ dz$$

$$p\,(t) = \frac{1}{\sqrt{2\pi n}\ 2^{n/2}\ \left\lfloor n/2 \right.}\ .\ \int_{0}^{\infty} e^{-\frac{z}{2}\left(1 + \frac{t^2}{n}\right)}\ .\ Z^{\frac{n+1}{2}-1}\ dz$$

$$p\,(t) = \frac{1}{\sqrt{2\pi n}\ 2^{n/2}\ \left\lfloor \dfrac{n}{2} \right.}\ \frac{\left\lfloor \dfrac{n+1}{2} \right.}{\left[\dfrac{1}{2}\left(1 + \dfrac{t^2}{n}\right)\right]^{\frac{n+1}{2}}}$$

$$p\,(t) = \frac{1}{\sqrt{n}}\ .\ \frac{\left\lfloor \dfrac{n+1}{2} \right.}{\sqrt{\pi}\ \left\lfloor \dfrac{n}{2} \right.}\ \frac{1}{\left(1 + \dfrac{t^2}{n}\right)^{\frac{n+1}{2}}} \qquad\qquad \left(\because \sqrt{\pi} = \left\lfloor \dfrac{1}{2} \right. \right)$$

$$p\,(t) = \frac{1}{\sqrt{n}}\ .\ \frac{1}{B\left(\dfrac{1}{2},\dfrac{n}{2}\right)}\ .\ \frac{1}{\left(1 + \dfrac{t^2}{n}\right)^{\frac{n+1}{2}}}\ ' \qquad -\infty < t < \infty$$

Thus, we can define a t-distribution with n degrees of freedom as a continuous r.v. having p.d.f.

$$f\,(t) = \frac{1}{\sqrt{n}}\ .\ \frac{1}{B\left(\dfrac{1}{2},\dfrac{n}{2}\right)}\ .\ \frac{1}{\left(1 + \dfrac{t^2}{n}\right)^{\frac{n+1}{2}}}\ ; \qquad -\infty < t < \infty$$

Note :

(1) It can be easily verified that $f(-t) = f(t)$, hence $f(t)$ is a symmetric function of t around 0.

(2) We symbolically write a r.v. follows t-distribution with n degrees of freedom as t_n.

2.2 Moments of t_n

(a)   **Mean** $= E(T) = \int_{-\infty}^{\infty} t\, f(t)\, dt = \int_{-\infty}^{\infty} \dfrac{1}{\sqrt{n}\, B\left(\dfrac{1}{2}, \dfrac{n}{2}\right)} \cdot \dfrac{t}{\left(1 + \dfrac{t^2}{n}\right)^{\frac{n+1}{2}}}\, dt$

Let, $g(t) = \dfrac{t}{\left(1 + \dfrac{t^2}{n}\right)^{\frac{n+1}{2}}}.$

Since, $g(-t) = -g(t)$, it is a odd function of t. Hence $\int_{-\infty}^{\infty} g(t)\, dt = 0$

$\therefore$ Mean $= E(t) = \mu_1' = 0.$

(b) Central moments of order 2r, (μ_{2r}) [Even ordered central moments] $(r = 1, 2 \ldots)$

Since, $\mu_1' = 0$, raw moments and central moments are same.

$\therefore$ $\mu_{2r} = E\left(t^{2r}\right) = \int_{-\infty}^{\infty} t^{2r}\, f(t)\, dt$

$= \int_{-\infty}^{\infty} \dfrac{1}{\sqrt{n}\, B\left(\dfrac{1}{2}, \dfrac{n}{2}\right)} \cdot \dfrac{t^{2r}}{\left(1 + \dfrac{t^2}{n}\right)^{\frac{n+1}{2}}}\, dt$

Since, the integrand is even function we write

$\mu_{2r} = \dfrac{1}{\sqrt{n}\, B\left(\dfrac{1}{2}, \dfrac{n}{2}\right)} \cdot 2\int_{0}^{\infty} \dfrac{t^{2r}}{\left(1 + \dfrac{t^2}{n}\right)^{\frac{n+1}{2}}}\, dt$

Making substitution $\dfrac{t^2}{n} = y$ we get,

$t = \sqrt{ny},\ 0 < y < \infty$ and $dt = \dfrac{\sqrt{n}}{2\sqrt{y}}\, dy$

$$\therefore \quad \mu_{2r} = \frac{1}{\sqrt{n}\, B\left(\frac{1}{2}, \frac{n}{2}\right)} \cdot \frac{2\sqrt{n}}{2} \int_0^\infty \frac{n^r y^r}{(1+y)^{\frac{n+1}{2}}} \frac{dy}{\sqrt{y}}$$

$$= \frac{n^r}{B\left(\frac{1}{2}, \frac{n}{2}\right)} \cdot \int_0^\infty \frac{y^{r+\frac{1}{2}-1}}{(1+y)^{\frac{n+1}{2}}}\, dy$$

$$= \frac{n^r}{B\left(\frac{1}{2}, \frac{n}{2}\right)} \cdot B\left(r+\frac{1}{2}, \frac{n}{2}-r\right) \qquad \left(\text{for } \frac{n}{2} > r\right); \; r = 1, 2, 3, \ldots$$

(i) If r = 1, then we get (for n > 2)

$$\text{Variance} = \mu_2 = n\, \frac{B\left(\frac{3}{2}, \frac{n}{2}-1\right)}{B\left(\frac{1}{2}, \frac{n}{2}\right)} = n\, \frac{\left\lceil \frac{3}{2}\right\rceil \left\lceil\frac{n}{2}-1\right\rceil}{\left\lceil\frac{1}{2}\right\rceil \left\lceil\frac{n}{2}\right\rceil} = n\, \frac{1/2}{(n/2-1)} = \frac{n}{n-2} \quad \text{(if } n > 2)$$

Comment : $\mu_2 > 1$

(ii) If r = 2, then for n > 4 we get,

$$\mu_4 = n^2\, \frac{B\left(\frac{5}{2}, \frac{n}{2}-2\right)}{B\left(\frac{1}{2}, \frac{n}{2}\right)} = n^2\, \frac{\left\lceil\frac{5}{2}\right\rceil \left\lceil\frac{n}{2}-2\right\rceil}{\left\lceil\frac{1}{2}\right\rceil \left\lceil\frac{n}{2}\right\rceil}$$

$$= n^2\, \frac{\frac{3}{2}\cdot\frac{1}{2}}{\left(\frac{n}{2}-1\right)\left(\frac{n}{2}-2\right)} = \frac{3\,n^2}{(n-2)\,(n-4)} \qquad \text{(if } n > 4)$$

(c) Central moments of order 2r + 1 (μ_{2r+1}) $(r = 1, 2, \ldots)$
[Odd ordered central moments]

$$\mu_{2r+1} = E\left(T^{2r+1}\right) = \int_{-\infty}^\infty t^{2r+1} f(t)\, dt$$

$$= \int_{-\infty}^\infty \frac{1}{\sqrt{n}\, B\left(\frac{1}{2}, \frac{n}{2}\right)} \frac{t^{2r+1}}{\left(1+\frac{t^2}{n}\right)^{\frac{n+1}{2}}}\, dt$$

Since, the integrand is odd function the integral is zero.

$$\therefore \quad \mu_{2r+1} = 0 \quad \text{for } r = 1, 2, 3, \ldots$$

Thus, all the moments of odd order are zero.

Therefore, first four central moments are.

$$\mu_1 = 0, \quad \mu_2 = \frac{n}{n-2}, \quad \mu_3 = 0, \quad \mu_4 = \frac{3n^2}{(n-2)(n-4)}$$

$\therefore$ Coefficients of skewness $\beta_1 = 0$, $\gamma_1 = 0$, hence the distribution is symmetric about mean (zero).

Also, the coefficient of kurtosis

$$\beta_2 = \frac{\mu_4}{\mu_2^2} = \frac{3n^2}{(n-2)(n-4)} \frac{(n-2)^2}{n^2} = \frac{3(n-2)}{n-4} = \frac{3(n-4+2)}{n-4} = 3 + \frac{6}{n-4} > 3$$

and $\qquad \gamma_2 = \beta_2 - 3 = \dfrac{6}{n-4} > 0$, hence the distribution is leptokurtic.

As $n \to \infty$, $\beta_2 \to 3$, hence the distribution is asymptotically mesokurtic.

Nature of probability curve

The probability curve is symmetric about 0 and bell shaped. Hence we get, mean = mode = median = 0. For large n the curve will resemble to that of standard normal curve.

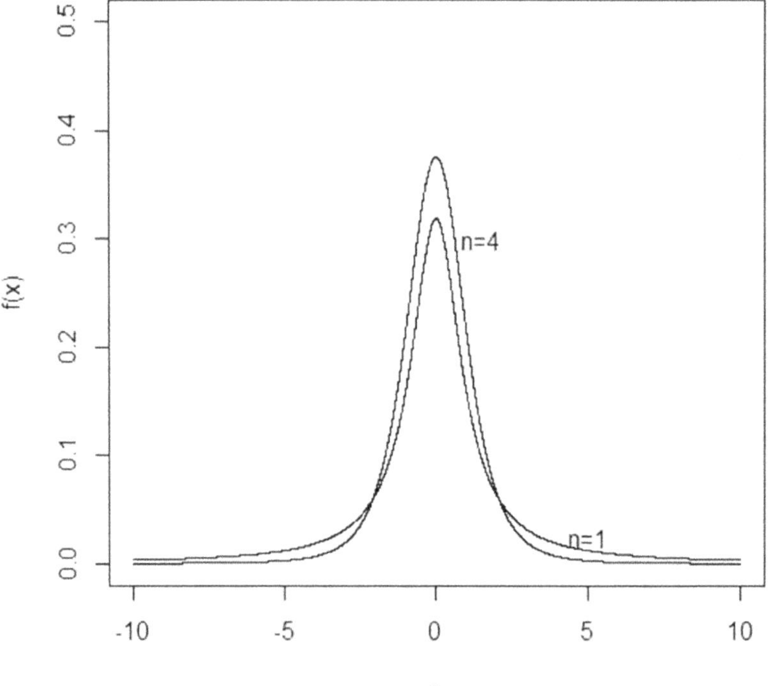

Fig. 2.1 : t-distribution probability curve using R software

Solved Examples

Example 2.1 : If a r.v. follows t-distribution with n degrees of freedom, then show that

the mean deviation about mean is $\sqrt{\dfrac{n}{\pi}}\ \dfrac{\left|\dfrac{n-1}{2}\right.}{\left|\dfrac{n}{2}\right.}$

Solution Note that mean of t-distribution is zero, hence mean deviation about mean is

$$E\,(|T|) \;=\; \int_{-\infty}^{\infty} |t| \cdot f\,(t)\,dt$$

Since the integrand is even function

$$E\,(|T|) \;=\; 2\int_{0}^{\infty} |t|\,f\,(t)\,dt$$

$$=\; 2.\int_{0}^{\infty} \frac{1}{\sqrt{n}\,B\!\left(\dfrac{1}{2},\dfrac{n}{2}\right)}\cdot\frac{t}{\left(1+\dfrac{t^2}{n}\right)^{\frac{n+1}{2}}}\,dt$$

put,$\quad \dfrac{t^2}{n} = y,\text{ hence } t = \sqrt{ny}\,,\ dt = \dfrac{\sqrt{n}\,dy}{2\sqrt{y}}$

$$\therefore\qquad E\,(|T|) \;=\; \frac{2}{\sqrt{n}\,B\!\left(\dfrac{1}{2},\dfrac{n}{2}\right)}\int_{0}^{\infty} \frac{\sqrt{ny}}{(1+y)^{\frac{n+1}{2}}}\cdot\frac{\sqrt{n}\,dy}{2\sqrt{y}}$$

$$=\; \frac{1}{B\!\left(\dfrac{1}{2},\dfrac{n}{2}\right)}\cdot\sqrt{n}\int_{0}^{\infty}\frac{1}{(1+y)^{\frac{n+1}{2}}}\,dy$$

$$=\; \sqrt{n}\ \frac{B\!\left(1,\dfrac{n-1}{2}\right)}{B\!\left(\dfrac{1}{2},\dfrac{n}{2}\right)} \;=\; \sqrt{\frac{n}{\pi}}\cdot\frac{\left|\dfrac{n-1}{2}\right.}{\left|\dfrac{n}{2}\right.}\quad [1 = y^0 = y^{1-1}]\qquad\qquad (\text{if } n > 1)$$

Example 2.2 : If μ_{2r} is central moment of order 2r of t-distribution with n d.f., then show that

$$\mu_{2r} \;=\; \frac{n\,(2r-1)}{n-2r}\,\mu_{2r-2}\ ;\quad r < \frac{n}{2}$$

Solution : Note that,

$$\mu_{2r} = E(t^{2r}) = n^r \frac{B\left(r + \frac{1}{2}, \frac{n}{2} - r\right)}{B\left(\frac{1}{2}, \frac{n}{2}\right)}$$

$$\therefore \quad \mu_{2r-2} = n^{r-1} \cdot \frac{B\left(r - 1 + \frac{1}{2}, \frac{n}{2} - r + 1\right)}{B\left(\frac{1}{2}, \frac{n}{2}\right)}$$

$$\therefore \quad \frac{\mu_{2r}}{\mu_{2r-2}} = n \cdot \frac{B\left(r + \frac{1}{2}, \frac{n}{2} - r\right)}{B\left(r - \frac{1}{2}, \frac{n}{2} - r + 1\right)}$$

$$= n \cdot \frac{\overline{\left|r + \frac{1}{2}\right.} \; \overline{\left|\frac{n}{2} - r\right.}}{\overline{\left|r - \frac{1}{2}\right.} \; \overline{\left|\frac{n}{2} - r + 1\right.}} = n \cdot \frac{\left(r - \frac{1}{2}\right)}{\left(\frac{n}{2} - r\right)} = n \cdot \frac{(2r - 1)}{(n - 2r)}$$

$$\therefore \quad \mu_{2r} = n \cdot \frac{(2r - 1)}{n - 2r} \cdot \mu_{2r-2}$$

Deduction : Note that $\mu_2 = \dfrac{n}{n - 2}$, the relation for $r = 2$ gives

$$\mu_4 = n \cdot \left(\frac{4 - 1}{n - 4}\right) \mu_2 = \frac{3n}{n - 4} \cdot \frac{n}{n - 2} = \frac{3n^2}{(n - 2)(n - 4)} \text{ if } n > 4.$$

2.3 Limiting Distribution of t_n

Statement : If a r.v. follows t-distribution with n degrees of freedom, then as $n \to \infty$ the probability distribution of T tends to N (0, 1).

Proof : We quote the results required in obtaining the limiting probability distribution.

Result (1) $\dfrac{\overline{\left|n + k\right.}}{\overline{\left|n\right.}} \approx n^k$ for large n

Result (2) $\displaystyle\lim_{n \to \infty} \left(1 + \frac{c}{n}\right)^n = e^c$, for c constant

$$\lim_{n \to \infty} f(t) = \lim_{n \to \infty} \frac{1}{\sqrt{n}\, B\left(\frac{1}{2}, \frac{n}{2}\right)} \times \frac{1}{\left(1 + \frac{t^2}{n}\right)^{\frac{n+1}{2}}}$$

$$= \lim_{n \to \infty} \frac{1}{\sqrt{n}} \cdot \frac{\left\lceil \dfrac{n+1}{2} \right.}{\left\lceil \dfrac{1}{2} \right. \left\lceil \dfrac{n}{2} \right.} \times \frac{1}{\left(1 + \frac{t^2}{n}\right)^{n/2}} \times \frac{1}{\left(1 + \frac{t^2}{n}\right)^{1/2}}$$

$$= \lim_{n \to \infty} \frac{1}{\sqrt{n}} \times \frac{1}{\sqrt{\pi}} \lim_{n \to \infty} \frac{\left\lceil \dfrac{n+1}{2} \right.}{\left\lceil \dfrac{n}{2} \right.} \times \lim_{n \to \infty} \frac{1}{\left[\left(1 + \frac{t^2}{n}\right)^n\right]^{1/2}} \times \frac{1}{\left(1 + \frac{t^2}{n}\right)^{1/2}}$$

$$= \lim_{n \to \infty} \frac{1}{\sqrt{\pi}} \frac{1}{\sqrt{n}} \cdot \left(\frac{n}{2}\right)^{\frac{1}{2}} \times \frac{1}{e^{t^2/2}}$$

$$= \frac{1}{\sqrt{2\pi}}\, e^{-t^2/2} \text{ which is the p.d.f. of N (0, 1).}$$

Hence the proof.

Note :

(1) For n = 1, we get

$$f(t) = \frac{1}{\pi} \cdot \frac{1}{1 + t^2} \qquad ; \qquad -\infty < t < \infty$$

which is called as Cauchy distribution.

(2) As $n \to \infty$, t-distribution tends to N (0, 1). However, the approximation holds good for n > 30.

Solved Examples

Example 2.3 : If $X_1, X_2 \ldots, X_n$ is a random sample from N (μ, σ^2), then show that

$Y = \dfrac{\overline{X} - \mu}{s \,|\, \sqrt{n}}$ follows t-distribution with n – 1 degrees of freedom where $s^2 = \dfrac{1}{n-1} \Sigma (X_i - \overline{X})^2$.

Solution : Note that $\overline{X}$ and $\dfrac{\Sigma (X_i - \overline{X})^2}{\sigma^2}$ are independent random variables. Moreover

$\overline{X} \to N (\mu, \sigma^2/n)$ and $\Sigma (X_i - \overline{X})^2/\sigma^2 \to \chi^2_{n-1}$.

Let, $U = \dfrac{\overline{X} - \mu}{\sigma/\sqrt{n}}$

Thus $U \to N (0, 1)$ and $V \to \chi^2_{n-1}$. Hence $Y = \dfrac{U}{\sqrt{V / (n-1)}} \to$ t-distribution with n – 1 d.f.

$$\therefore \quad Y = \frac{\bar{X} - \mu}{\sigma/\sqrt{n}} \times \frac{1}{\sqrt{\dfrac{\Sigma \, (X_i - \bar{X})^2}{(n-1)\,\sigma^2}}} = \frac{(\bar{X} - \mu)\,\sqrt{n}}{\sigma} \times \frac{\sigma}{s}$$

$$Y = \frac{\bar{X} - \mu}{s/\sqrt{n}} \quad \text{follows t-distribution with } (n-1) \text{ d.f.}$$

Example 2.4 If t follows Student's t-distribution with 15 d.f. find

    (i)   c such that $P \, (-c \le t \le c) = 0.8$

    (ii) k such that $P \, (t^2 \ge k) = 0.05$

    (iii) a such that $P \, (t \le a) = 0.1$

Solution : From statistical table we get $P \, (|t| \ge c)$

    (i) $P \, (-c \le t \le c) = P \, (|t| \le c) = 0.8$

$\therefore$ $P \, (|t| > c) = 0.2,$

From statistical table we get, $P \, (|t_{15}| > 1.341) = 0.2$

$\therefore$ $C = 1.341$

    (ii) $P \, (t^2 \ge k) = P \, (|t_{15}| \ge \sqrt{k}) = 0.05$

From statistical table we get,

     $P \, (|t_{15}| > 2.131) = 0.05$, therefore $\sqrt{k} = 2.131$

$\therefore$ $k = 4.5412$

    (iii) $P \, (t \le a) = 0.1 \quad a < 0$

         $P \, (t \ge -a) = 0.1.$ Due to symmetry.

$\therefore$ $P \, [|t| \ge -a] = 0.2$

From statistical tables,

      $P \, (|T_{15}| > 1.341) = 0.2$

$\therefore$ $a = -1.341$

COMPUTATION OF PROBABILITIES FOR t-DISTRIBUTION

Compute $P \, [t_{15} > 1.341]$ using :

    (A) Statistical tables   (B) MS-Excel   (C) R-software.

(A)   Statistical table gives $P \, (|t_{15}| > |1.34|) = 0.2$

             $P \, (t_{15} > 1.341) + P \, (t_{15} < -1.341) = 0.2$

Due to symmetry $P \, (t_{15} > 1.34) = P \, (t_{15} < -1.341) = 0.1.$

(B)   Using MS-EXCEL we find $P \, (t_{15} > 1.341)$ as follows :

Step 1 : Click on Insert on the Excel sheet.

Step 2 : Clock on $\boxed{fx}$ i.e. function.

Step 3 : Select function $\boxed{TDIST}$.

After completing these steps a box as shown below will appear on the screen.

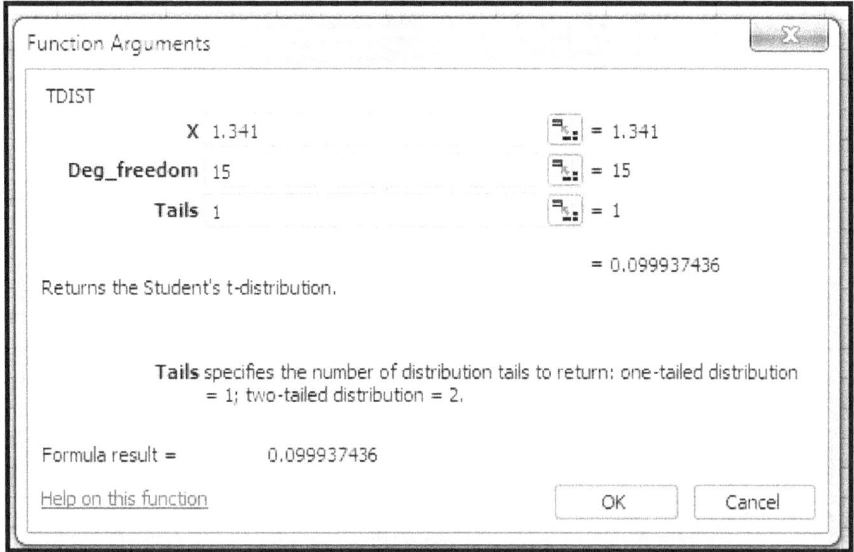

Fig. 2.2 (a)

Enter the value of X as 1.341, De-freedom as 15. As we want to find one tailed probability enter Tail as 1. Whenever we want to compute two tailed probabilities enter number of Tail as 2.

Step 4 : Click on $\boxed{OK}$.

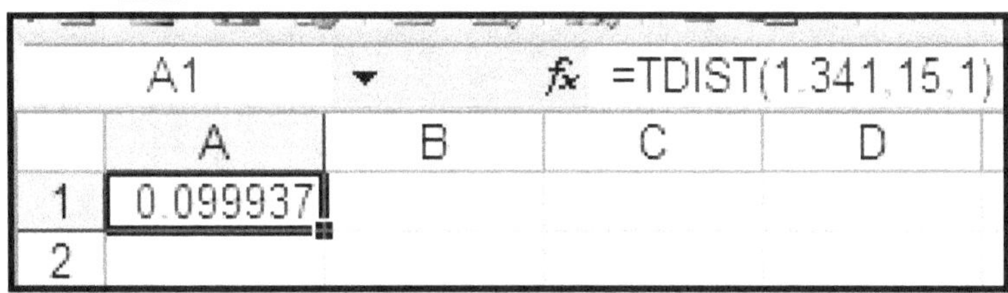

Fig. 2.2 (b)

We get the answer as 0.099937 ≈ 0.01.

Remark : In case we want to obtain P [|t| > 1.341] using MS-EXCEL, we have to enter 2 in the dialog box against $\boxed{Tails}$. Then click on $\boxed{OK}$ we get answer on the screen as 0.19987.

Note : The above probability can be directly obtained using MS-EXCEL Command = TDIST (1.341, 15, 1).

(C)   Using R-software : We find probabilities of events of t.

R-software supports following commands to compute probabilities :

(a) $\boxed{pt\ (x,\ n)}$: It gives P (X ≤ x) where X has t-distribution with n d.f.

Example : To compute (i) P $(t_{10} > 3.9)$ (ii) P $(t_{12} < 6)$.

(i)   P $(t_{10} > 3.9) = 1 - $ P $(t_{10} \leq 3.0)$

∴ R command is $\boxed{1 - pt\ (3.9,\ 10)}$ which gives answer 0.0014805.

(ii) P $(t_{12} < 6)$ is computed by command $\boxed{pt\ (6,\ 12)}$ which gives answer 0.999969.

(b) qt (p, n) : It gives ordinate below which t_n r.v. has probability P.

Example : (i) To find C such that P $(t_{15} \leq C) = 0.1$.

R command is $\boxed{qt\ (0.1,\ 15)}$ which gives answer as C = $- 1.340606$.

(ii) To find C such that P $(t_{17} \geq C) = 0.4$.

P $(t_{17} \geq C) = 1 - $ P $(t_{17} \leq C)$.

∴   P $(t_{17} \leq C) = 0.6$.

∴   R-command is $\boxed{qt\ (0.6,\ 17)}$ which gives answer of C = 0.257347.

Points to Remember

1. Suppose U follows N (0, 1) and V follows chi-square distribution with n degrees of freedom. If U and V are independent random variables then t $= \dfrac{U}{\sqrt{V/n}}$ follows t-distribution with n degrees of freedom.

2. The p.d.f. of t-distribution with n degrees of freedom is

$$f(t) \ = \frac{1}{\sqrt{n}} \cdot \frac{1}{B\left(\dfrac{1}{2}, \dfrac{n}{2}\right)} \cdot \frac{1}{\left(1 + \dfrac{t^n}{n}\right)^{\frac{n+1}{2}}} \ ; \ -\infty < t < \infty$$

3. If T $\rightarrow t_n$ with n degrees of freedom then

(i)   E(T) = mean = 0, Var (T) = $\mu_2 = \dfrac{n}{n-2}$.

(ii) All odd ordered central moments are zero i.e. $\mu_{2r+1} = 0$ for r = 1, 2, 3 … .

(iii) All even ordered central moments are given by

$$\mu_{2r} = \frac{n^r}{B\left(\dfrac{1}{2}, \dfrac{n}{2}\right)} \ B\left(r + \frac{1}{2}, \frac{n}{2} - r\right) \left(\text{for } \frac{n}{2} > r\right) \ r = 1, 2 \dots$$

(iv) $\mu_4 = \dfrac{3n^2}{(n-2)\,(n-4)}$, $\beta_1 = 0$, $\gamma_1 = 0$, $\beta_2 = 3 + \dfrac{6}{n-4}$, $\gamma_2 = \dfrac{6}{n-4}$.

(v) The mean deviation about mean $= \sqrt{\dfrac{n}{\pi}} \; \dfrac{\Gamma\left(\dfrac{n-1}{2}\right)}{\Gamma\left(\dfrac{n}{2}\right)}$.

(vi) As $n \to \infty$ the probability distribution of T tends to N (0, 1).

4. If μ_{2r} is central moment of order 2r of t-distribution with n d.f. then

$$\mu_{2r} = \frac{n(2r-1)}{n-2r} \; \mu_{2r-2}; \quad r < \frac{n}{2}$$

5. If $X_1, X_2, \ldots, X_n$ is a random sample from N (μ, σ^2) then Y $= \dfrac{\bar{X}-\mu}{s/\sqrt{n}}$ follows t-distribution

with $(n-1)$ degrees of freedom where $s^2 = \dfrac{1}{n-1} \Sigma \, (X_i - \bar{X})^2$.

Exercise 2 (A)

(1) Define Student's t-distribution and derive its p.d.f.
(P.U. May 2011, 2002, April 2004)

(2) Show that the moments of odd order of t-distribution are zero.

(3) Define students t-distribution. Find the $(2r)^{th}$ moment of t-distribution with n d.f. (2r < n). Hence find its variance. **(P.U. April 98, Oct. 98, Nov. 2004)**

(4) Show that, t-distribution is symmetric around zero. **(P.U. Oct. 2006)**

(5) With usual notation, show that for t-distribution with n d.f.

$$\mu_{2r} = \frac{n\,(2r-1)}{n-2r} \; \mu_{2r-2}; \quad 2r < n$$

(6) (a) If t follows Student's t-distribution with n d.f., obtain the distribution of

(i) $X = t^2$ (ii) $Y = \dfrac{1}{1 + \dfrac{t^2}{n}}$

(b) If X follows t-distribution with n d.f. Find the probability distribution of X^2. Identify it, also draw its probability curve. **(P.U. April 98)**

(7) Find the mean deviation about mean of t-distribution. with n d.f.
(P.U. Oct. 2012, April 98, Oct. 98, May 2001, 2005)

(8) If t is a r.v. with Student's t-distribution with n d.f. such that $\dfrac{\mu_6}{\mu_4} = 6$, find n.

(9) If $X_1, X_2, \ldots, X_{10}$ be i.i.d. N (0, 1) variates, then obtain the probability distribution of

$$Y = \frac{3\,X_{10}}{\sqrt{\displaystyle\sum_{1}^{9} X_i^2}}.$$

(10) Let $U \to N(0, 1)$, $V \to \chi_n^2$. If U and V are independent r.v. s. find
$$P\left(\frac{U + \sqrt{V/n}}{\sqrt{V/n}} \le 1\right).$$

(11) Suppose $X_1, X_2, ..., X_n$ is a random sample from $N(\mu, \sigma^2)$ and $Y_1, Y_2, ... Y_m$ is another random sample from $N(\mu, \sigma^2)$. Determine the probability distribution of
$$Y = \sqrt{\frac{n - m}{n + m}}\,(\bar{X} - \bar{Y}) \bigg/ \sqrt{\frac{\Sigma(X_i - \bar{X})^2 + \Sigma(Y_i - \bar{Y})^2}{n + m - 2}}.$$

(12) If t_n is a r.v. with Student's t-distribution with n d.f. find

    (i)   c such that $P(-c \le t_{12} \le c) = 0.8$

    (ii)   k such that $P(t_{16}^2 \ge k) = 0.05$

    (iii)   a such that $P(t_{10} \le a) = 0.2$.

(13) If X_1, X_2, X_3 are i.i.d. $N(0, 1)$ variates and $Y = \dfrac{\sqrt{2}\,X_3}{\sqrt{X_1^2 + X_2^2}}$, find C such that

   $P(Y \le C) = 0.995$.

(14) Show that a r.v. having student's t-distribution with n d.f. tends to $N(0, 1)$ as n tends to ∞.

(15) If a random variable T follows t-distribution with n degrees of freedom, find the $(2r)^{th}$ order central moment of T. Hence find its variance.   **(P.U. Oct. 2011)**

(16) Derive the mean and variance of t-distribution with n degrees of freedom.
          (P.U. May 2013)

(17) If t_n follows Student's t-distribution with n d.f. find
(i) $P(|t_{10}| > 1.812)$ (ii) $P(|t_8| \le 2.306)$
(iii) $P(t_{26} < 0.531)$ (iv) $P(t_{30} > 1.697)$.   **(P.U. 2004)**

(18) Suppose X is a r.v. assumed to follow student t-distribution with 12 d.f. Find the constant C such that $P(X \le C) = 0.10$.   **(P.U. May 2002)**

(19) If $X \to \chi_n^2$, $Y \to \chi_n^2$ such that X and Y are independent, determine the probability distribution of $Z = \dfrac{\sqrt{n}\,(X - Y)}{2\sqrt{XY}}$.

(20) If X follows t-distribution with n d.f. show that $Y = X^2$ follow F-distribution with $(1, n)$ d.f.   **(P.U. 2003)**

(21) Let X have t-distribution with 8 degrees of freedom. Find a and b such that
$P(X^2 \le a) = 0.01$ and $P(X^2 \ge b) = 0.01$ **(P.U. 1997)**

(22) Let $(X_i, i = 1, 2, 3, 4, 5)$ be i.i.d. $N(0, 4)$ variates

   Find $P\left(\dfrac{2X_5}{\sqrt{X_1^2 + X_2^2 + X_3^2 + X_4^2}} \ge 1.19\right)$.   **(P.U. 1987)**

Exercise 2 (B)

Choose correct alternative out of (a) to (d) for following questions.

(1) If X follows t-distribution with 5 d.f. then Var (X) is

(a) $\dfrac{3}{5}$　　　(b) $\dfrac{5}{3}$　　　(c) $\dfrac{4}{3}$　　　(d) $\dfrac{3}{4}$

(2) If X is a random variable having t-distribution with n d.f. such that $\dfrac{\mu_6}{\mu_4}$ = 7 then n is equal to

(a) 20　　　(b) 22　　　(c) 24　　　(d) 21

(3) Let t be random variable following student's t-distribution with n d.f. $\dfrac{\mu_6}{\mu_4}$ = 8.

(a) $\dfrac{8}{7}$　　　(b) $\dfrac{7}{8}$　　　(c) $\dfrac{16}{15}$　　　(d) $\dfrac{15}{16}$

(4) Suppose $X_1, X_2, ..., X_5$ are i.i.d. N (0, 1) variates. Then the probability distribution of

$$\frac{2X_5}{\sqrt{\sum_1^4 X_i^2}} \text{ is}$$

(a) χ_4^2　　　(b) χ_5^2　　　(c) t_4　　　(d) t_5

(5) If $X_1, X_2, ..., X_{10}$ are i.i.d. N (0, 1) variates then the probability distribution of

$$\frac{3\bar{X}}{\sqrt{\sum_{i=1}^{10} (X_i - \bar{X})^2}} \text{ is}$$

(a) t_{10}　　　(b) χ_{10}^2　　　(c) χ_9^2　　　(d) t_9

(6) If $X_1, X_2, X_3, ..., X_8$ are i.i.d N (0, 1) variates then $\dfrac{X_1 - X_2 + X_3 - X_4}{\sqrt{X_5^2 + X_6^2 + X_7^2 + X_8^2}}$ has

(a) t-distribution with 8 d.f.　　　(b) t-distribution with 4 d.f.
(c) χ^2-distribution with 8 d.f.　　　(d) χ^2-distribution with 4 d.f.

(7) Let X_1, X_2, X_3 be i.i.d. N (0, 1) variates and Y = $\dfrac{\sqrt{2} \, X_3}{\sqrt{X_1^2 + X_2^2}}$. Hence E (Y) is

(a) $\sqrt{2}$　　　(b) 1　　　(c) 2　　　(d) 0

(8) Suppose X_1, X_2, X_3, X_4 and X_5 independent variables such that each of X_1, X_2, X_3, X_4 has N (0, 1) distribution while X_5 follows N (0, 4) distribution. Hence median of the distribution of $\dfrac{X_5}{\sqrt{X_1^2 + X_2^2 + X_3^2 + X_4^2}}$ is

(a) 0　　　(b) 2　　　(c) 4　　　(d) 3

(9) If X_1 and X_2 are i.i.d. N (0, 1) and $Y = \dfrac{X_1 + X_2}{|X_1 - X_2|}$ then the distribution of y is

     (a) t_2 (b) t_1 (c) χ_2^2 (d) χ_1^2

(10) If $X_1, X_2 \ldots X_{16}$ is a random sample from N (μ, σ^2) and $\sum\limits_{i=1}^{16} (X_i - \bar{X})^2 = 375$ then the distribution of $\dfrac{4\,(\bar{X} - \mu)}{5}$ will be

     (a) t_{16} (b) t_{14} (c) t_{13} (d) t_{15}

(11) If X follows t-distribution with 6 d.f. then the fourth central moment of the distribution μ_4 is

     (a) $\dfrac{27}{4}$ (b) $\dfrac{25}{4}$ (c) $\dfrac{27}{2}$ (d) $\dfrac{25}{2}$

(12) The coefficient of kurtosis γ_2 for t-distribution with 10 d.f. is

     (a) 1   (b) 2   (c) 4   (d) 0

(13) If X follows t-distribution with 5 d.f. then variance of X is

     (a) $\dfrac{3}{5}$ (b) $\dfrac{4}{3}$ (c) $\dfrac{3}{4}$ (d) $\dfrac{5}{3}$

(14) If T follows student's t-distribution with 10 d.f. and P $[T \le K] = 0.2$ then K is equal to

     (a) $-\,0.879$ (b) 0.879   (c) $-\,1.372$ (d) 1.372

(15) If $X \sim > t_n$ then X^2 follows :

     (a) χ_n^2 (b) t_{2n} (c) $F_{1,\,n}$ (d) $F_{n,\,1}$

(16) If $X \sim > t_n$ then E(X) is

     (a) n   (b) $\dfrac{n}{n-2}$ (c) 0   (d) 2n

(17) If X follows t-distribution with 6 degrees of freedom then the second ordered central moment of X is

     (a) 2/3   (b) 3/2   (c) 3/4   (d) 6

Exercise 2 (C)

State whether the following statements are True or False.

(1) Suppose X is a r.v. which follows t-distribution with n d.f. Then Var (X) is an integer only when n = 3 or 4.

(2) If a r.v. has t-distribution with n d.f. then the central moments of order 2r (μ_{2r}) are

given by $\mu_{2r} = \dfrac{n^r}{B\left(\dfrac{n}{2}, \dfrac{1}{2}\right)} \; B\left(r + \dfrac{1}{2}, \dfrac{n}{2} - r\right)$ for $\dfrac{n}{2} > r$.

(3) If X follows t-distribution then Var (X) cannot be equal to 5.

(4) If a r.v. has t-distribution with n d.f. then the central moments of order 2r (μ_{2r}) are

given by $\mu_{2r} = \dfrac{n^r}{B\left(\dfrac{1}{2}, \dfrac{n}{2}\right)} \; B\left(r + \dfrac{1}{2}, \dfrac{n}{2} - r\right)$ for $\dfrac{n}{2} < r$.

(5) If a r.v. follow t-distribution with 2 d.f. then mean deviation about mean is $\sqrt{2}$.

(6) If μ_{2r} is central moment of order 2r of t-distribution with n d.f. then

$$\mu_6 = \frac{n}{n-6} \, \mu_4.$$

(7) If μ_{2r} is central moment of order 2r of t-distribution with n d.f. then

$$\mu_{2r} = \frac{n\,(2r - 1)}{n - 2r} \, \mu_{2r-2} \text{ for } r > \frac{n}{2}.$$

(8) The t-distribution is symmetric about 1.

(9) The t-distribution is asymptotically mesokurtic.

(10) All even ordered central moments for t-distribution are zero.

(11) For t-distribution; A.M. = Median = Mode = 0.

(12) Students t-distribution is asymptotically leptokurtic

Answers of Exercise 2 (A)

(6) (i) $f(x) = \dfrac{1}{\sqrt{n}} \dfrac{1}{\beta\left(\dfrac{1}{2}, \dfrac{n}{2}\right)} \cdot \dfrac{1}{\sqrt{x}} \cdot \dfrac{1}{\left(1 + \dfrac{x}{n}\right)^{\frac{n+1}{2}}}$; $x > 0$

(ii) $f(y) = \dfrac{1}{\beta\left(\dfrac{1}{2}, \dfrac{n}{2}\right)} \cdot y^{\frac{n}{2}-1} (1-y)^{-1/2}.$

(8) n = 36 (9) t_9 (10) 0.5 (11) $t_{n_1 + n_2 - 2}$ (12) (i) 1.356 (ii) 4.4944 (iii) 0.26

(13) 9.925 (15) (i) 0.1 (ii) 0.95 (iii) 0.025 (iv) 0.05 (17) t_n

(21) $a = \dfrac{1}{5982}$, b = 11.26 (22) 0.15

Answers of Exercise 2 (B)

(1) b	(2) d	(3) a	(4) c
(5) d	(6) b	(7) d	(8) a
(9) b	(10) d	(11) c	(12) a
(13) d	(14) c	(15) c	(16) c
(17) b			

Answers of Exercise 2 (C)

(1) True	(2) True	(3) True	(4) False
(5) True	(6) False	(7) False	(8) False
(9) True	(10) False	(11) True	(12) False

Chapter **3**...

Snedecore's F-Distribution

George Waddel Snedecor (October 20, 1881 – February 15, 1974) was anAmerican mathematician and statistician. He contributed to the foundations of analysis of variance, data analysis, experimental design, and statistical methodology. Snedecor's F distribution and the George W. Snedecor Awards of the American Statistical Association are named after him.

George Waddel Snedecor

Contents ...

Key Words : F-distribution.

Objectives :

(1) To derive the probability density function of F-distribution using two independent chi-square variates.

(2) To study the properties of F-distribution.

(3) To understand the interrelationship among T, F, χ^2 variables.

3.0 Introduction

Like Student's t-distribution, Snedecor's F-distribution is useful in statistical inference. Such as tests of hypothesis for equality of variances, equality of means of three or more groups, non-linearity of regression. It is defined as distribution of ratio of two independent chi-square random variables.

3.1 P.D.F. of F-Distribution

Definition : Suppose U and V are independent chi-square random variables with n_1 and n_2 degrees of freedom respectively then $F = \dfrac{U/n_1}{V/n_2}$ is said to follow Snedencor's F-distribution with parameters n_1 and n_2. We symbolically write it as $F_{(n_1 \cdot n_2)}$.

Derivation of p.d.f. of $F_{(n_1 \cdot n_2)}$:

Since F is a transformation of U and V, we find its p.d.f. as follows :

(i)   Let $F = \dfrac{U/n_1}{V/n_2}$ and $Z = V$ Clearly, $0 < F < \infty$ and $0 < Z < \infty$.

(ii)   We get, $U = \dfrac{n_1}{n_2} F Z$ and $V = Z$.

Hence the Jacobian of transformation

$$
J = \frac{\partial (U, V)}{\partial (F, Z)} =
\begin{vmatrix}
\dfrac{\partial U}{\partial F} & \dfrac{\partial U}{\partial Z} \\[2mm]
\dfrac{\partial V}{\partial F} & \dfrac{\partial V}{\partial Z}
\end{vmatrix}
=
\begin{vmatrix}
\dfrac{n_1}{n_2} Z & \dfrac{n_1}{n_2} F \\[2mm]
0 & 1
\end{vmatrix}
= \frac{n_1}{n_2} Z
$$

(iii)   Joint p.d.f. of (F, Z) = Joint p.d.f. of (U, V) . $|J|$

$$g (F, Z) = f (U, V) . |J|$$

$$= f_1 (U) . f_2 (V) |J| \quad (\because U \text{ and } V \text{ are independent})$$

$$= \frac{\left(\dfrac{1}{2}\right)^{\frac{n_1}{2}}}{\left|\dfrac{n_1}{2}\right.} e^{-\frac{U}{2}} . U^{\frac{n_1}{2} - 1} . \frac{\left(\dfrac{1}{2}\right)^{\frac{n_2}{2}}}{\left|\dfrac{n_2}{2}\right.} . e^{-\frac{V}{2}} V^{\frac{n_2}{2} - 1} . |J|$$

$$= \frac{1}{2^{\frac{n_1 + n_2}{2}} \left|\dfrac{n_1}{2}\right. \left|\dfrac{n_2}{2}\right.} . e^{-\left(\frac{(U + V)}{2}\right)} . U^{\frac{n_1}{2} - 1} . V^{\frac{n_2}{2} - 1} |J|$$

$$= C . e^{-\frac{1}{2}\left(\frac{n_1}{n_2} FZ + Z\right)} \left(\frac{n_1}{n_2} fz\right)^{\frac{n_1}{2} - 1} . Z^{\frac{n_2}{2} - 1} \frac{n_1}{n_2} Z$$

$$\text{where, } C = \frac{1}{2^{\frac{n_1 + n_2}{2}} \left|\dfrac{n_1}{2}\right. \left|\dfrac{n_2}{2}\right.}$$

$$\therefore \quad g(f, z) = C. \left(\frac{n_1}{n_2}\right)^{\frac{n_1}{2}} . F^{\frac{n_1}{2} - 1} Z^{\frac{n_1 + n_2}{2} - 1} e^{-\frac{1}{2}\left(1 + \frac{n_1}{n_2} F\right) Z}$$

(iv)　To get marginal p.d.f. of F we integrate g (F, Z) over the range of Z.

$$P(f) = C. \left(\frac{n_1}{n_2}\right)^{\frac{n_1}{2}} . F^{\frac{n_1}{2} - 1} \int_0^{\infty} e^{-\frac{1}{2}\left(1 + \frac{n_1}{n_2} F\right) Z} . Z^{\frac{n_1 + n_2}{2} - 1} . dz$$

$$= C. \left(\frac{n_1}{n_2}\right)^{\frac{n_1}{2}} . f^{\frac{n_1}{2} - 1} \frac{\left|\frac{n_1 + n_2}{2}\right.}{\left[\frac{1}{2}\left(1 + \frac{n_1}{n_2} f\right)\right]^{\frac{n_1 + n_2}{2}}}$$

$$= \left(\frac{n_1}{n_2}\right)^{\frac{n_1}{2}} . \frac{\left|\frac{n_1 + n_2}{2}\right.}{\left|\frac{n_1}{2}\right. \left|\frac{n_2}{2}\right.} \times \frac{F^{\frac{n_1}{2} - 1}}{\left(1 + \frac{n_1}{n_2} F\right)^{\frac{n_1 + n_2}{2}}}$$

$$\therefore \quad P(f) = \frac{\left(\frac{n_1}{n_2}\right)^{\frac{n_1}{2}}}{B\left(\frac{n_1}{2}, \frac{n_2}{2}\right)} \times \frac{f^{\frac{n_1}{2} - 1}}{\left(1 + \frac{n_1}{n_2} f\right)^{\frac{n_1 + n_2}{2}}} , \quad F > 0$$

Note : Symbolically we write a r.v. F follows F distribution with n_1 and n_2 d.f. as $F \to F_{(n_1, n_2)}$.

3.2 Mean, Variance, Moments of $F_{(n_1, n_2)}$.

(a)　Mean　　$E(F) = \int_0^{\infty} f. P(f) \, df$

$$= \int_0^{\infty} \frac{\left(\frac{n_1}{n_2}\right)^{\frac{n_1}{2}}}{B\left(\frac{n_1}{2}, \frac{n_2}{2}\right)} . F . \frac{F^{\frac{n_1}{2} - 1}}{\left(1 + \frac{n_1}{n_2} F\right)^{\frac{n_1 + n_2}{2}}} \, dF$$

Substituting $x = \frac{n_1}{n_2} F.$, $dF = \frac{n_2}{n_1} dx$

$$\therefore \quad E(F) = \frac{1}{B\left(\frac{n_1}{2},\frac{n_2}{2}\right)} \cdot \int_0^\infty \frac{x^{\frac{n_1}{2}}}{(1+x)^{\frac{n_1+n_2}{2}}} \cdot \frac{n_2}{n_1}\, dx$$

$$= \frac{n_2}{n_1} \cdot \frac{1}{B\left(\frac{n_1}{2},\frac{n_2}{2}\right)} \times B\left(\frac{n_1}{2}+1,\frac{n_2}{2}-1\right) \qquad \text{if } \frac{n_2}{2} > 1$$

$$= \frac{n_2}{n_1} \cdot \frac{\left\lfloor\frac{n_1}{2}+1\right.\;\left\lfloor\frac{n_2}{2}-1\right.}{\left\lfloor\frac{n_1+n_2}{2}\right.} \times \frac{\left\lfloor\frac{n_1+n_2}{2}\right.}{\left\lfloor\frac{n_1}{2}\right.\;\left\lfloor\frac{n_2}{2}\right.} \qquad \text{if } n_2 > 2$$

$$= \frac{n_2}{n_1} \cdot \frac{n_1/2}{(n_2/2-1)} = \frac{n_2}{n_1} \cdot \frac{n_1}{(n_2-2)} = \frac{n_2}{n_2-2} \qquad \text{if } n_2 > 2$$

Note :

(i)   Mean $= \dfrac{n_2}{n_2-2}$ which is independent of the other parameter n_1.

(ii)   Mean $= \dfrac{n_2}{n_2-2} > 1.$

(iii)   We can verify here that $E\left(\dfrac{X}{Y}\right) \neq \dfrac{E(X)}{E(Y)}$. If U and V are independent chi-square r.v.s. with n_1 and n_2 d.f. respectively then $\dfrac{U/n_1}{V/n_2}$ follows $F(n_1, n_2)$.

$$\therefore \quad E\left(\frac{U/n_1}{V/n_2}\right) = \frac{n_2}{n_2-2} \text{ where as } \frac{E(U/n_1)}{E(V/n_2)} = 1$$

Thus if $X = U/n_1$ and $Y = V/n_2$ then, $E\left(\dfrac{X}{Y}\right) \neq \dfrac{E(X)}{E(Y)} = 1$

(b) r^{th} Raw Moment

$$\therefore \quad \mu_r' = E\left(F^r\right) = \int_0^\infty F^r \cdot P(F) \cdot dF = \int_0^\infty \frac{\left(\frac{n_1}{n_2}\right)^{\frac{n_1}{2}}}{B\left(\frac{n_1}{2},\frac{n_2}{2}\right)} \cdot F^r \frac{F^{\frac{n_1}{2}-1}}{\left(1+\frac{n_1}{n_2}F\right)^{\frac{n_1+n_2}{2}}}\, dF$$

Substituting $\dfrac{n_1}{n_2} F = x$ $\therefore F = \dfrac{n_2}{n_1} = x$ we get $dF = \dfrac{n_2}{n_1} dx$

$$\therefore \quad \mu_r' = \frac{1}{B\left(\dfrac{n_1}{2}, \dfrac{n_2}{2}\right)} \cdot \left(\frac{n_1}{n_2}\right)^{\frac{n_1}{2}} \cdot \int_0^\infty \frac{\left(\dfrac{n_2}{n_1} x\right)^{\frac{n_1}{2} + r - 1}}{(1 + x)^{\frac{n_1 + n_2}{2}}} \cdot \frac{n_2}{n_1} dx$$

$$= \left(\frac{n_2}{n_1}\right)^r \cdot \frac{B\left(\dfrac{n_1}{2} + r, \dfrac{n_2}{2} - r\right)}{B\left(\dfrac{n_1}{2}, \dfrac{n_2}{2}\right)} \cdot \qquad \text{if } \frac{n_2}{2} > r$$

In particular for r = 1 we get

$$\mu_1' = \frac{n_2}{n_1} \cdot \frac{B\left(\dfrac{n_1}{2} + 1, \dfrac{n_2}{2} - 1\right)}{B\left(\dfrac{n_1}{2}, \dfrac{n_2}{2}\right)} = \frac{n_2}{n_2 - 2}$$

For r = 2, we get

$$\mu_2' = \left(\frac{n_2}{n_1}\right)^2 \cdot \frac{B\left(\dfrac{n_1}{2} + 2, \dfrac{n_2}{2} - 2\right)}{B\left(\dfrac{n_1}{2}, \dfrac{n_2}{2}\right)} \qquad \text{if } \frac{n_2}{2} > 2$$

$$= \left(\frac{n_2}{n_1}\right)^2 \cdot \frac{\overline{\dfrac{n_1}{2} + 2} \; \overline{\dfrac{n_2}{2} - 2}}{\overline{\dfrac{n_1}{2}} \; \overline{\dfrac{n_2}{2}}} = \left(\frac{n_2}{n_1}\right)^2 \frac{\left(\dfrac{n_1}{2} + 1\right)\left(\dfrac{n_1}{2}\right)}{\left(\dfrac{n_1}{2} - 1\right)\left(\dfrac{n_2}{2} - 2\right)} \quad \text{if } n_2 > 4$$

$$= \left(\frac{n_2}{n_1}\right)^2 \frac{n_1 (n_1 + 2)}{(n_2 - 2)(n_2 - 4)} = \frac{n_2^2 (n_1 + 2)}{n_1 (n_2 - 2)(n_2 - 4)}$$

Variance $= \mu_2' - \left(\mu_1'\right)^2 \qquad \text{if } n_2 > 4.$

$$= \frac{n_2^2 (n_1 + 2)}{n_1 (n_2 - 2)(n_2 - 4)} - \left(\frac{n_2}{n_2 - 2}\right)^2 = \frac{n_2^2}{n_2 - 2}\left[\frac{n_1 + 2}{n_1 (n_2 - 4)} - \frac{1}{n_2 - 2}\right]$$

$$= \frac{n_2^2}{n_2 - 2}\left[\frac{n_1 n_2 - 2 n_2 - 2 n_1 - 4 - n_1 n_2 + 4 n_1}{n_1 (n_2 - 2)(n_2 - 4)}\right] = \frac{2 n_2^2 (n_1 + n_2 - 2)}{n_1 (n_2 - 2)^2 (n_2 - 4)} \quad \text{if } n_2 > 4.$$

3.3 Mode of $F_{(n_1, n_2)}$

M is mode of the distribution if it is a point of maxima of p.d.f. P (F). It can be obtained by solving P' (F) = 0, such that P" (F) < 0.

$$P\,(F)\ =\ \frac{\left(\dfrac{n_1}{n_2}\right)^{\frac{n_1}{2}}}{B\left(\dfrac{n_1}{2},\dfrac{n_2}{2}\right)}\ \cdot\ \frac{F^{\frac{n_1}{2}-1}}{\left(1+\dfrac{n_1}{n_2}F\right)^{\frac{n_1+n_2}{2}}}$$

$$P\,(F)\ =\ C\cdot\frac{F^{\frac{n_1}{2}-1}}{\left(1+\dfrac{n_1}{n_2}F\right)^{\frac{n_1+n_2}{2}}},\quad \text{where,}\ \ C=\frac{\left(\dfrac{n_1}{n_2}\right)^{\frac{n_1}{2}}}{B\left(\dfrac{n_1}{2},\dfrac{n_2}{2}\right)}$$

Taking logarithms on both sides

$$\therefore\quad \log P\,(F) = \log C + \left(\frac{n_1}{2}-1\right)\log F - \left(\frac{n_1+n_2}{2}\right)\log\left(1+\frac{n_1}{n_2}F\right)$$

Differentiating both the sides w.r.t. F, we get

$$\frac{P'\,(F)}{P\,(F)}\ =\ 0 + \frac{\frac{n_1}{2}-1}{F} - \left(\frac{n_1+n_2}{2}\right)\frac{\frac{n_1}{n_2}}{1+\frac{n_1}{n_2}F}$$

$$\therefore\quad P'\,(F)\ =\ \left\{\frac{n_1-2}{2\,F} - \frac{n_1\,(n_1+n_2)}{2\,n_2}\times\frac{n_2}{n_2+n_1\,F}\right\} P\,(F) \qquad\qquad \ldots (1)$$

$$\therefore\quad P'\,(F)\ =\ 0,\ \text{gives}$$

$$\therefore\quad \frac{n_1-2}{2\,F}\ =\ \frac{n_1\,(n_1+n_2)}{2\,(n_2+n_1\,F)}$$

$$\therefore\quad (n_2+n_1\,F)\,(n_1-2) = n_1\,(n_1+n_2)\,F$$

$$\therefore\quad n_1\,(n_1-2)\,F - n_1\,n_2\,F = n_1^2\,F - n_2\,(n_1-2)$$

$$\therefore\quad F\ =\ \frac{n_2}{n_2+2}\cdot\frac{n_1-2}{n_1}\ ;\ \text{if}\ n_1 > 2$$

Differentiating equation (1) w.r.t. F we can verify

$$P''(F) < 0 \text{ at } \frac{n_2(n_1 - 2)}{n_1(n_2 + 2)}$$

Hence the mode is $\left(\dfrac{n_2}{n_2 + 2}\right)\left(\dfrac{n_1 - 2}{n_1}\right)$.

Remark : Note that $\dfrac{n_2}{n_2 + 2} < 1$ and $\dfrac{n_1 - 2}{n_1} < 1$. Hence mode is less than 1.

3.4 Nature of Probability Curve

Note that mean is $\dfrac{n_2}{n_2 - 2}$ (if $n_2 > 2$).

which is greater than 1. However mode is smaller than 1. Hence Mean – Mode > 0. Therefore

Karl Pearson's coefficient of skewness $= \dfrac{\text{Mean} - \text{Mode}}{\text{S.D.}} > 0$ for $n_1 > 2$ and $n_2 > 2$. It indicates that the probability curve has positive skewness. As $n_1 \to \infty$, $n_2 \to \infty$ the curve becomes symmetric.

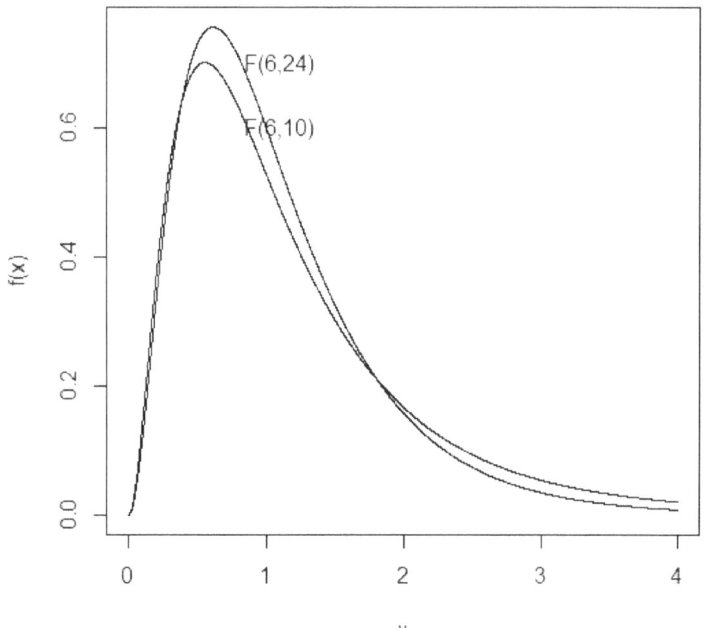

Fig. 3.1 : F-distribution probability curve

Distribution	Mean	Mode
$F_{(6, 24)}$	1.0909	0.6154
$F_{(6, 10)}$	1.25	0.5556

Result (1) : If X follows F-distribution with n_1, n_2 d.f., then $\dfrac{1}{X}$ follows F distribution with n_2, n_1 d.f.

Solution : Let $Y = \dfrac{1}{X}$, therefore $Y > 0$.

$$X = \frac{1}{Y}, \; J = \frac{dx}{dy} = -\frac{1}{Y^2}$$

$\therefore$ p.d.f. Y is given by,

   $g(y) = f(x) \, |J|$ where, $f(x)$ is p.d.f. of X

$\therefore$ $g(y) = \dfrac{\left(\dfrac{n_1}{n_2}\right)^{\frac{n_1}{2}}}{B\left(\dfrac{n_1}{2}, \dfrac{n_2}{2}\right)} \cdot \dfrac{x^{\frac{n_1}{2}-1}}{\left(1+\dfrac{n_1}{n_2}x\right)^{\frac{n_1+n_2}{2}}} \cdot |J|$

$= \dfrac{\left(\dfrac{n_1}{n_2}\right)^{\frac{n_1}{2}}}{B\left(\dfrac{n_1}{2}, \dfrac{n_2}{2}\right)} \cdot \dfrac{\left(\dfrac{1}{y}\right)^{\frac{n_1}{2}-1}}{\left(1+\dfrac{n_1}{n_2 y}\right)^{\frac{n_1+n_2}{2}}} \cdot \dfrac{1}{y^2}$

$= \dfrac{\left(\dfrac{n_1}{n_2}\right)^{\frac{n_1}{2}}}{B\left(\dfrac{n_1}{2}, \dfrac{n_2}{2}\right)} \cdot \left(\dfrac{1}{y}\right)^{\frac{n_1}{2}+1} \left[\dfrac{\dfrac{n_2 y}{n_1}}{\dfrac{n_2 y}{n_1}+1}\right]^{\frac{n_1+n_2}{2}}$ $\left[\begin{array}{l}\text{Multiplying Numerator and} \\ \text{Denominator by} \\ \left(\dfrac{n_2}{n_1}y\right)^{\frac{n_1+n_2}{2}}\end{array}\right]$

$= \dfrac{\left(\dfrac{n_1}{n_2}\right)^{\frac{n_1}{2}} \cdot \left(\dfrac{n_2}{n_1}\right)^{\frac{n_1+n_2}{2}}}{B\left(\dfrac{n_1}{2}, \dfrac{n_2}{2}\right)} \cdot \dfrac{y^{\frac{n_2}{2}-1}}{\left(1+\dfrac{n_2 y}{n_1}\right)^{\frac{n_1+n_2}{2}}}$

$= \dfrac{\left(\dfrac{n_2}{n_1}\right)^{\frac{n_2}{2}}}{B\left(\dfrac{n_2}{2}, \dfrac{n_1}{2}\right)} \cdot \dfrac{y^{\frac{n_2}{2}-1}}{\left(1+\dfrac{n_2}{n_1}y\right)^{\frac{n_1+n_2}{2}}} \; ; \; y > 0$ $\left[\because B\left(\dfrac{n_1}{2}, \dfrac{n_2}{2}\right) = B\left(\dfrac{n_2}{2}, \dfrac{n_1}{2}\right)\right]$

which is the p.d.f. of $F_{(n_2,\, n_1)}$

Remark : Thus reciprocal of F variate also follows F distribution (with d.f. in reverse order).

3.5 Median and Quartiles of $f_{(n, n)}$

Median : If F is a r.v. with F distribution with (n, n) d.f. then the median of F is 1.

By definition m is median if

$$P\left(F_{(n, n)} \geq m\right) = P\left(F_{(n, n)} \leq m\right) = \frac{1}{2}. \qquad \text{... (1)}$$

$$\therefore \qquad P\left(\frac{1}{F_{(n, n)}} \leq \frac{1}{m}\right) = \frac{1}{2}$$

Since $\dfrac{1}{F_{(n, n)}}$ also follows $F_{(n, n)}$ we get

$$P\left(F_{(n, n)} \leq \frac{1}{m}\right) = \frac{1}{2}$$

$$\therefore \qquad P\left(F_{(n, n)} \geq \frac{1}{m}\right) = \frac{1}{2} \qquad \text{... (2)}$$

Thus from equation (1) and (2) we get

$$m = \frac{1}{m} \text{ i. e. } m^2 = 1$$

$$\therefore \qquad m = 1 \text{ is the median.} \qquad (\because m > 0)$$

Result 2 : If F follows $F_{(n, n)}$ the first and the third quartiles are reciprocals of each other.

Proof : Let Q_1 and Q_3 be the first and the third quartiles respectively.

By definition $P\left(F_{(n, n)} \leq Q_1\right) = \dfrac{1}{4}$

$$\therefore \qquad P\left(\frac{1}{F_{(n, n)}} \geq \frac{1}{Q_1}\right) = \frac{1}{4}$$

$$\therefore \qquad P\left(F_{(n, n)} \geq \frac{1}{Q_1}\right) = \frac{1}{4} \qquad \left(\because \frac{1}{F_{(n, n)}} \to F_{(n, n)}\right)$$

$$\therefore \qquad P\left(F_{(n, n)} \leq \frac{1}{Q_1}\right) = \frac{3}{4} \qquad \text{... (1)}$$

However Q_3 is defined as

$$P\left(F_{(n, n)} \leq Q_3\right) = \frac{3}{4} \qquad \text{... (2)}$$

Therefore comparing equation (1) and (2), we get

$$\frac{1}{Q_1} = Q_3 \quad \text{ or } Q_1 = \frac{1}{Q_3} \quad \text{ or } Q_1 Q_3 = 1$$

Result 3 : If $X \to F(n_1, n_2)$ and $Y \to F(n_2, n_1)$ then show that $P(X \geq a) + P\left(Y \geq \dfrac{1}{a}\right) = 1$.

Proof : Let, $P(X \geq a) = P$ ... (1)

$$\therefore \quad P\left(\frac{1}{X} \leq \frac{1}{a}\right) = P$$

$$\therefore \quad P\left(Y \leq \frac{1}{a}\right) = P \qquad \left(\because \frac{1}{X} \to F_{(n_2,\, n_1)}\right)$$

$$\therefore \quad P\left(Y \geq \frac{1}{a}\right) = 1 - P \qquad \text{... (2)}$$

$\therefore$ From equation (1) and (2) we get

$$P(X \geq a) + P\left(Y \geq \frac{1}{a}\right) = 1.$$

Note : The above result equivalent to

$$P(X \geq a) = P\left(Y \leq \frac{1}{a}\right)$$

i.e. $P\left(F(n_1, n_2) \geq a\right) = P\left(F(n_2, n_1) \leq \frac{1}{a}\right)$

Corollary : If $F_{(n_1,\, n_2,\, \alpha)}$ is a point such that

$$P\left[F_{(n_1, n_2)} \geq F_{(n_1, n_2, \alpha)}\right] = \alpha$$

then show that $F_{(n_1,\, n_2,\, \alpha)} = \dfrac{1}{F_{(n_1,\, n_2,\, 1-\alpha)}}$

Proof : $P\left[F_{(n_1, n_2)} \geq F_{(n_1, n_2, \alpha)}\right] = \alpha$

$$\therefore \quad P\left[\frac{1}{F_{(n_1, n_2)}} \leq \frac{1}{F_{(n_1, n_2, \alpha)}}\right] = \alpha$$

$$\therefore \quad P\left[F_{(n_2, n_1)} \leq \frac{1}{F_{(n_1, n_2, \alpha)}}\right] = 1 - \alpha \qquad \text{... (1)}$$

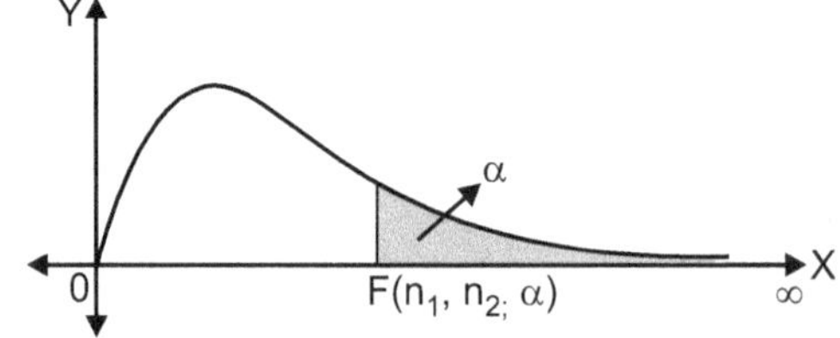

Fig. 3.2

However by definition

$$P\left[F_{(n_2, n_1)} \geq F_{(n_2, n_1, 1-\alpha)}\right] = 1 - \alpha \qquad \text{... (2)}$$

$\therefore$ From equation (1) and (2) we get $F_{(n_1,\, n_2,\, \alpha)} = \dfrac{1}{F_{(n_2,\, n_1,\, 1-\alpha)}}$

3.6 Inter Relationship Among T, F, χ^2 Variables

(A) Relation between T and F.

Note that if U and V are independent r.v.s., $U \to N(0, 1)$, $V \to \chi^2_n$ then we get $t = \dfrac{U}{\sqrt{V/n}}$

as t variate with n d.f. Clearly $t^2 = \dfrac{U^2}{V/n} = \dfrac{\chi^2_1 V_1}{\chi^2_2/n} = F_{(1, n)}$. Thus we get if $t \to t_n$ then $t^2 \to F_{(1, n)}$.

Result : It $t \to t_n$, then $t^2 \to F(1, n)$.

Proof : Let $t \to t_n$, and $Y = t^2$. We derive distribution function of Y and differentiate it to get p.d.f.

$$
\begin{aligned}
G(y) &= P(Y \le y) = P(t^2 \le y) \\
&= P(-\sqrt{y} \le t \le \sqrt{y}) \\
&= P(t \le \sqrt{y}) - P(t \le -\sqrt{y}) \\
&= F(\sqrt{y}) - F(-\sqrt{y})
\end{aligned}
$$

F (.) is a distribution function of t.

$$
\therefore \quad \text{p.d.f. of } Y = g(y) = \frac{d}{dy} G(y)
$$

$$
= \frac{d}{dy} F(\sqrt{y}) - \frac{d}{dy} F(-\sqrt{y})
$$

$$
= \frac{f(\sqrt{y})}{2\sqrt{y}} - \frac{f(-\sqrt{y})}{2\sqrt{y}} = \frac{f(\sqrt{y}) + f(-\sqrt{y})}{2\sqrt{y}} \qquad \text{where } f(.) \text{ is a p.d.f. of } t
$$

$$
= \frac{f(\sqrt{y})}{\sqrt{y}} \qquad\qquad (\because f(-\sqrt{y}) = f(\sqrt{y}))
$$

$$
\therefore \quad g(y) = \frac{1}{\sqrt{y}} \cdot \frac{1}{\sqrt{n}\, B\left(\frac{1}{2}, \frac{n}{2}\right)} \times \frac{1}{\left(1 + \frac{y}{n}\right)^{\frac{n+1}{2}}}
$$

$$
= \frac{\left(\frac{1}{n}\right)^{\frac{1}{2}}}{B\left(\frac{1}{2}, \frac{n}{2}\right)} \cdot \frac{y^{\frac{1}{2} - 1}}{\left(1 + \frac{y}{n}\right)^{\frac{n+1}{2}}} \qquad\qquad (\because y > 0)
$$

which is p.d.f. of $F_{(1, n)}$

(B) Relation between F and χ^2 :

(i) F-distribution is defined as distribution of the ratio of independent chi-square variates divided by their corresponding degrees of freedom.

(ii)   If F follows $F_{(n_1, n_2)}$ then probability distribution of $n_{1/F}$ tends to chi-square probability distribution with n_1 degrees of freedom as $n_2 \to \infty$.

Proof :   Let $Y = n_1 F$, hence $J = \dfrac{dF}{dY} = \dfrac{1}{n_1}$

$\therefore$ p.d.f. Y = p.d.f. of F $|J|$

$$g(y) = \frac{1}{n_1} \cdot \frac{\left(\dfrac{n_1}{n_2}\right)^{\frac{n_1}{2}}}{B\left(\dfrac{n_1}{2}, \dfrac{n_2}{2}\right)} \cdot \frac{F^{\frac{n_1}{2} - 1}}{\left(1 + \dfrac{n_1 F}{n_2}\right)^{\frac{n_1 + n_2}{2}}}$$

$$= \frac{\left(\dfrac{n_1}{n_2}\right)^{\frac{n_1}{2}}}{B\left(\dfrac{n_1}{2}, \dfrac{n_2}{2}\right)} \cdot \frac{\left(\dfrac{1}{n_1}\right)^{\frac{n_1}{2}} \cdot y^{\frac{n_1}{2} - 1}}{\left(1 + \dfrac{y}{n_2}\right)^{\frac{n_1 + n_2}{2}}}$$

$$\lim_{n_2 \to \infty} g(y) = \lim_{n_2 \to \infty} \frac{n_2^{\frac{-n_1}{2}} \dfrac{\left\lceil \dfrac{n_1 + n_2}{2} \right.}{ }}{\left\lceil \dfrac{n_1}{2} \right. \left\lceil \dfrac{n_2}{2} \right.} \cdot \frac{y^{\frac{n_1}{2} - 1}}{\left(1 + \dfrac{y}{n_2}\right)^{\frac{n_2}{2}} \left(1 + \dfrac{y}{n_2}\right)^{\frac{n_1}{2}}}$$

Note that for large n, $\dfrac{\left\lceil n + k \right.}{\left\lceil n \right.} \approx n^k$, k fixed $\therefore$ $\lim_{n_2 \to \infty} \dfrac{\Gamma\left(\dfrac{n_1 + n_2}{2}\right)}{\Gamma\left(\dfrac{n_2}{2}\right)} = \left(\dfrac{n_2}{2}\right)^{\frac{n_1}{2}}$

and $\lim_{a \to \infty} \left(1 + \dfrac{k}{a}\right)^a = e^k$

$$\therefore \lim_{n_2 \to \infty} g(y) = \lim_{n_2 \to \infty} \frac{n_2^{-\frac{n_1}{2}} \left(\dfrac{n_2}{2}\right)^{\frac{n_1}{2}}}{\left\lceil \dfrac{n_1}{2} \right.} \cdot \frac{y^{\frac{n_1}{2} - 1}}{\left(1 + \dfrac{y}{n_2}\right)^{\frac{n_2}{2}} \left(1 + \dfrac{y}{n_2}\right)^{\frac{n_1}{2}}}$$

$$= \frac{\left(\dfrac{1}{2}\right)^{\frac{n_1}{2}}}{\left\lceil \dfrac{n_1}{2} \right.} \cdot \frac{y^{\frac{n_1}{2} - 1}}{e^{y/2}} = \frac{\left(\dfrac{1}{2}\right)^{\frac{n_1}{2}}}{\left\lceil \dfrac{n_1}{2} \right.} \cdot e^{-y/2} \, y^{\frac{n_1}{2} - 1} \; ; y > 0$$

It is p.d.f. of χ_n^2 r.v. with n_1 d.f.

Solved Examples

Example 3.1 : If X follows $F_{(2, n)}$, then show that

$$P(X \geq k) = \left(1 + \frac{2k}{n}\right)^{-\frac{n}{2}}, \quad k > 0.$$ Also find $\lim_{n \to \infty} P(X \geq k)$ and compare it with

probability of that of a χ_2^2 variate. Thus verify $\lim_{n \to \infty} P(X \geq k) = P\left(\lim_{n \to \infty} X \geq k\right)$.

Solution : (i) Since $X \to F_{(2, n)}$, its p.d.f. will be

$$f(x) = \frac{2}{n} \cdot \frac{1}{B\left(1, \frac{n}{2}\right)} \times \frac{1}{\left(1 + \frac{2x}{n}\right)^{\left(\frac{n+2}{2}\right)}} = \frac{2}{n} \cdot \frac{\left\lfloor \frac{n}{2} + 1 \right.}{\left\lfloor \frac{n}{2} \right.} \left(1 + \frac{2x}{n}\right)^{-\left(\frac{n+2}{2}\right)}$$

$$= \left(1 + \frac{2x}{n}\right)^{-\left(\frac{n+2}{2}\right)}$$

$$\therefore \quad P(X \geq k) = \int_{k}^{\infty} \left(1 + \frac{2x}{n}\right)^{-\frac{n+2}{2}} dx = \left[\frac{\left(1 + \frac{2x}{n}\right)^{\frac{n}{2}}}{-\frac{n}{2}} \times \frac{2}{n}\right]_{k}^{\infty}$$

$$= -\lim_{x \to \infty} \left(1 + \frac{2x}{n}\right)^{-\frac{n}{2}} + \left(1 + \frac{2k}{n}\right)^{-\frac{n}{2}} = 0 + \left(1 + \frac{2k}{n}\right)^{-\frac{n}{2}}$$

$$\lim_{n \to \infty} P(X \geq k) = \lim_{n \to \infty} \left(1 + \frac{2k}{n}\right)^{-\frac{n}{2}} = e^{-k}$$

(ii) If $X \to F_{(n_1, n_2)}$ then as $n_2 \to \infty$, $n_1 F \to \chi_{n_1}^2$

Here $X \to F_{(2, n)}$, therefore as $n \to \infty$, $2X \to \chi_2^2$

$$\therefore \quad P\left(\lim_{n \to \infty} X \geq k\right) = P\left(\lim_{n \to \infty} 2X \geq 2k\right)$$

$$= P(\chi_2^2 \geq 2k) = \int_{2k}^{\infty} \frac{1}{2} e^{-y/2} \, dy$$

$$= \left[-e^{-y/2}\right]_{2k}^{\infty} = -e^{-\infty} + e^{-\frac{2k}{2}} = e^{-k}$$

Note : Thus from (i) and (ii) we have verified that

$$\lim_{n \to \infty} P\,(X \geq k) = P\left(\lim_{n \to \infty} X \geq k\right) = e^{-k}$$

Example 3.2 : Suppose $X_1, X_2, ..., X_{25}$ is a random sample from N $(\mu_1, 10)$ and $Y_1, Y_2, ... Y_{21}$ is a random sample from N $(\mu_2, 15)$. If $S_1^2 = \dfrac{1}{24} \Sigma\,(X_i - \bar{X})^2$ and $S_2^2 = \dfrac{1}{20} \Sigma(Y_i - \bar{Y})^2$, then find $P\left(\dfrac{S_1^2}{S_2^2} > 1.3867\right)$.

Solution : If X_i 's are i.i.d. N (μ, σ^2) then $\displaystyle\sum_{1}^{25} (X_i - \bar{X})^2/\sigma^2$ follows χ^2 distribution with 24 d.f.

$$\therefore \quad \sum_{1}^{25} (X_i - \bar{X})^2/10 \to \chi_{24}^2 \text{ and } \sum_{2}^{21} (Y_i - Y)^2/15 \to \chi_{20}^2$$

$$\therefore \quad \text{i.e. } 24\,S_1^2/10 \to \chi_{24}^2 \text{ and } 20S_2^2/15 \to \chi_{20}^2$$

$$\therefore \quad \frac{S_1^2/10}{S_2^2/15} \to F_{(24,\,20)}$$

$$\therefore \quad P\,(S_1^2/S_2^2 > 1.3867) = P\left(\frac{S_1^2/10}{S_2^2/15} > 1.3867 \times \frac{15}{10}\right)$$

$$= P\,(F_{(24,\,20)} > 2.08) = 0.05$$

COMPUTATION OF PROBABILITIES OF f-DISTRIBUTION USING MS-EXCEL

Find P $[F_{24,\,20} > 2.08]$ using :

(A) Statistical tables, (B) MS-Excel, (C) R-software.

Solution : (A) Using statistical table, we get P $(X > 2.08) = 0.05$ if $X \to F\,(24, 20)$.

(B) Using MS-EXCEL.

Step 1 : Click on $\boxed{\text{Insert}}$ on MS-EXCEL sheet.

Step 2 : Click on $\boxed{\text{fx}}$.

Step 3 : Selection function $\boxed{\text{FDIST}}$.

Then the box as given below will appear on the screen.

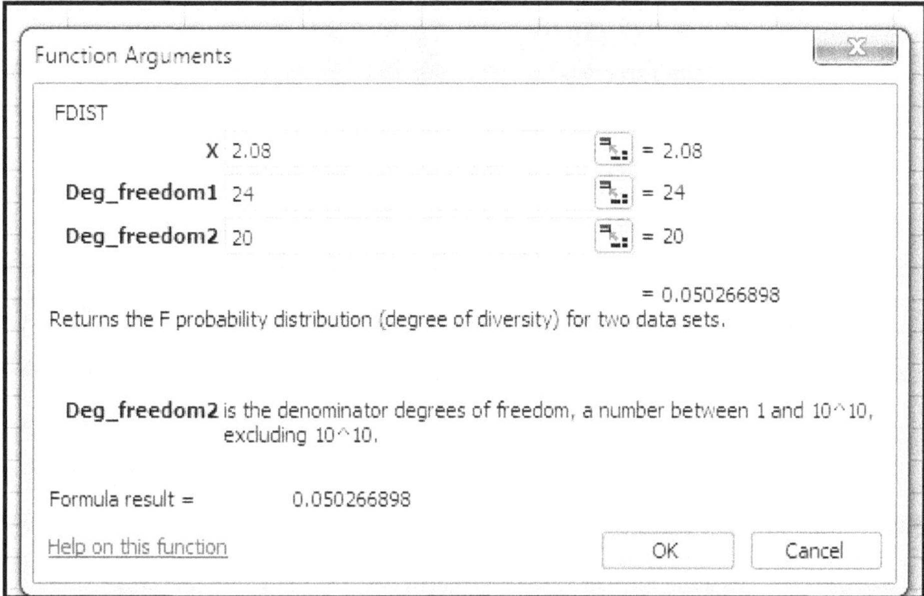

Fig. 3.3 (a)

Enter the value of X as 2.08, Deg_freedom1 as 24 and Deg_freedom2 as 20.

Step 4 : Click on $\boxed{\text{OK}}$ we get the answer as 0.05067.

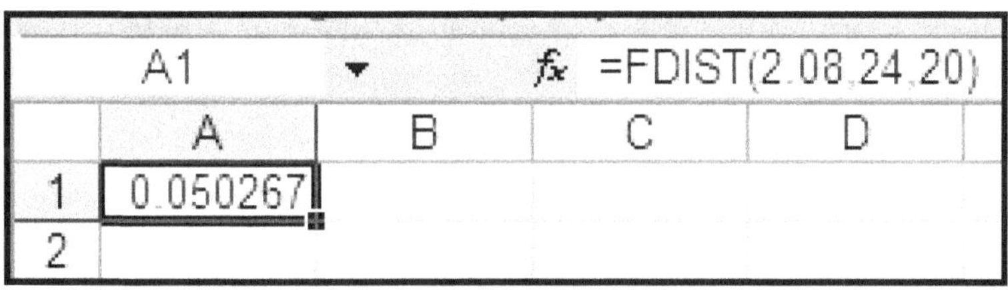

Fig. 3.3 (b)

Note : The above probability can be directly obtained using MS-EXCEL Command $\boxed{= \text{FIDIST}(2.08, 24, 20)}$. The command gives probability of right tail.

(C) Using R-software : R-software provides following commands to compute probabilities of F-distribution.

(a)   pf (x, n_1, n_2) : It gives P (X ≤ x) where X has F-distribution with n_1 and n_2 d.f.

(i)   To compute P ($F_{5,\,8} \le 9$).

R-command is $\boxed{\text{pf (9, 5, 8)}}$ which gives answer 0.996137.

(ii) To compute P ($F_{7,\,6} \ge 4$).

P ($F_{7,\,6} \ge 4$) = 1 − P ($F_{7,\,6} \le 4$)

∴   R-command is $\boxed{1 - \text{pf (4, 7, 6)}}$ which gives answer 0.05580012.

(b)   qf (p, n_1, n_2) : It gives ordinate below which F_{n_1, n_2} r.v. has probability p.

(i)   To find k such that p ($F_{4, 7} \leq k$) = 0.5.

R-command is $\boxed{\text{qf (0.5, 4, 7)}}$ which gives answer 0.9261931 as value of k.

(ii)   To find k such that P ($F_{8, 12} \geq k$) = 0.8

$\therefore$ P ($F_{8, 12} \geq k$) = 1 − P ($F_{8, 12} \leq k$)

R-command is $\boxed{\text{qt (0.2, 8, 12)}}$ which gives answer 0.547919 as value of k.

Points to Remember

1.   If χ_1^2 and χ_2^2 are two independent chi-square random variables with n_1 and n_2 degrees of freedom respectively then the ratio $\dfrac{\chi_1^2 / n_1}{\chi_2^2 / n_2}$ follows F-distribution with n_1 and n_2 degrees of freedom.

2.   The probability density function (p.d.f.) of F-distribution with n_1 and n_2 degrees of freedom is given by

$$P(F) = \frac{\left(\dfrac{n_1}{n_2}\right)^{\frac{n_1}{2}}}{B\left(\dfrac{n_1}{2}, \dfrac{n_2}{2}\right)} \cdot \frac{F^{\frac{n_1}{2} - 1}}{\left(1 + \dfrac{n_1}{n_2} F\right)^{\frac{n_1 + n_2}{2}}} \; ; \;\; F > 0$$

3.   If F follows F-distribution with n_1 and n_2 degrees of freedom then

(i)   E(F) = mean = $\dfrac{n_2}{n_2 - 2}$ > 1 which independent of parameter n_1.

(ii)   The rth raw moment is given by

$$\mu_r' = \left(\frac{n_2}{n_1}\right)^r \frac{B\left(\dfrac{n_1}{2} + r, \dfrac{n_2}{2} - r\right)}{B\left(\dfrac{n_1}{2}, \dfrac{n_2}{2}\right)} \quad \text{if } \frac{n_2}{2} > r.$$

In particular $\mu_2' = \dfrac{n_2^2 (n_1 + 2)}{n_1 (n_2 - 2)(n_2 - 4)}$ if $n_2 > 4$

(iii) Var (F) = $\dfrac{2n_2^2 (n_1 + n_2 - 2)}{n_1 (n_1 - 2)^2 (n_2 - 4)}$ if $n_2 > 4$

(iv) Mode = $\left(\dfrac{n_2}{n_2 + 2}\right)\left(\dfrac{n_1 - 2}{n_1}\right)$ if $n_1 > 2$. It is always less than 1.

(v)   Mean − Mode > 0 $\therefore$ The density curve is positively skewed.

4. If F is a random variable with (n, n) d.f. then (i) median $= 1$, (ii) $Q_1 Q_3 = 1$.

5. If $X \to F(n_1, n_2)$ and $Y \to F(n_2, n_1)$ then $P(X \ge a) + P\left(Y \ge \dfrac{1}{a}\right) = 1$ where a is constant.

6. If X follows F distribution with n_1, n_2 degrees of freedom then $\dfrac{1}{X}$ follows F-distribution with n_2, n_1 degrees of freedom. Thus reciprocal of F variate also follows F-distribution with d.f. in reverse order.

7. If $t \to t_n$ then $t^2 \to F_{(1, n)}$.

8. If F follows $F_{(n_1, n_2)}$ then $n_1 F$ tends to chi-square probability distribution with n_1 degrees of freedom as $n_2 \to \infty$.

9. F-distribution with parameters n, n is symmetric.

Exercise 3 (A)

(1) Define Snedecor's F-distribution with n_1 and n_2 d.f. and derive its p.d.f.

(P.U. Nov. 2004, Oct. 2005, 2006)

(2) Derive r^{th} raw moment of F-distribution with n_1 and n_2 d.f. Hence find mean and variance of the distribution.

(P.U. May 2011, April 99, 2004; May 2001, May 2002)

(3) Show that median of $F_{(n, n)}$ is unity.

(4) Show that the product of the first and third quartiles of $F(n, n)$ is unity.

(P.U. May 2013, Oct. 98)

(5) (a) Derive mode of $F_{(n_1, n_2)}$ and show that it is smaller than mean. Hence comment on the symmetry of the distribution.   **(P.U. April 98)**

(b) Show that mode of F-distribution with n_1 and n_2 d.f. is $\dfrac{n_2 (n_1 - 2)}{n_1 (n_2 + 2)}$, $n_1 > 2$.

(P.U. Oct. 2012)

(6) State the inter-relations among normal, chi-square, t and F-distribution.

(P.U. Oct. 2012, April 99, May 2002)

(7) If $X \to$ t-distribution with n d.f. show that $X^2 \to F_{(1, n)}$.   **(P.U. Oct. 98, May 2003)**

(8) If X and Y are independent variables such that $X \to G(\alpha, \lambda_1)$ and $Y \to G(\alpha, \lambda_2)$ then find the probability distribution of $(\lambda_2 X)/(\lambda_1 Y)$.

(9) If X and Y are i.i.d. exponential with mean θ, then find the distribution of X/Y.

(10) If $X \to F_{(n_1, n_2)}$ find the probability distributions of (i) $\dfrac{1}{X}$ **(P.U. May 2013)**

(ii) $\dfrac{1}{1 + \dfrac{n_1}{n_2} X}$.

(11) If $F \to F_{(n_1, n_2)}$ then find the probability distribution of $n_1 F$ as $n_2 \to \infty$.

(12) If X is a continuous r.v. with p.d.f.

$$f(x) = \frac{1}{B\left(\frac{1}{2}, \frac{n-2}{2}\right)} (1 - x^2)^{\frac{n-4}{2}} \; ; -1 \le x \le 1, n > 2$$

then find the probability distribution of $\frac{1+X}{1-X}$.

(13) Let X and Y be F variates with (m, n) and (n, m) d.f. respectively. Show that :
$P(X \ge a) + P(Y \ge 1/a) = 1.$ **(P.U. 1990, April 2000)**

(14) Let X and Y be two i.i.d. gamma variates with parameters $(1, \lambda)$. Show that

$$P\left(\frac{X}{Y} \le 1\right) = \frac{1}{2}.$$ **(P.U. 1987)**

(15) Let X and Y have F distribution with (3, 6) and (6, 3) d.f. respectively. Find the value
of $P(X < 10) + P(Y \le 0.1)$.   **(P.U. 1987)**

(16) Let X have Snedecor's F distribution with 2 and 8 d.f. If $P(X \le a) = 0.01$ and
$P(X \ge b) = 0.99$, find a and b.   **(P.U. 1989)**

(17) A r.v. X has p.d.f.

$$f(x) = k \frac{x^{\frac{n_1}{2} - 1}}{\left(1 + \frac{n_1}{n_2} x\right)^{\frac{n_1 + n_2}{2}}} ; \qquad\qquad x > 0, n_1, n_2 > 0.$$

Find the constant k and state the probability distribution of $\frac{1}{X}$.   **(P.U. 1991)**

(18) Let X_1, X_2, X_3 be independent normal variates with $E(X_i) = i$ and var $(X_i) = i^2$,
$i = 1, 2, 3$. Using X_1, X_2, X_3 construct a statistic that has F distribution with 1 and
2 d.f.   **(P.U. 1996)**

(19) If X_1, X_2, X_3, X_4 are i.i.d. N (0, 1) variates, find α, such that

$$P\left(\frac{3 X_4^2}{X_1^2 + X_2^2 + X_3^2} \le \alpha\right) = 0.01$$ **(P.U. 1991)**

(20) With usual notations prove that $F_{n_1, n_2, \alpha} = F_{n_2, n_1, (1-\alpha)}^{-1}.$ **(P.U. May 2003)**

(21) Let X; $i = 1, 2, ..., 5$ be independent standard normal variates. Identify the

distribution of $Y = \dfrac{4X_1^2}{X_2^2 + X_3^2 + X_4^2 + X_5^2}$. Hence find the value of C such that

$P(Y \ge C) = 0.01.$ **(May 2003)**

(22) Let $X_i \sim$ i.i.d. N (0, 1) where, i = 1, 2, ... variates. Identify the distribution of

(i) $\dfrac{X_1^2 + X_2^2 + X_3^2}{X_4^2 + X_5^2 + X_6^2}$ **(P.U. 97)** (ii) $\dfrac{X_1}{\sqrt{\dfrac{X_2^2 + X_6^2}{2}}}$ **(P.U. 1999)**

(23) Let X have t-distribution with 8 d.f. Find a and b such that
P $(X^2 \le a)$ = 0.01 and P $(X^2 \ge b)$ = 0.01. **(P.U. 1997)**

Exercise 3 (B)

Choose correct alternative out of (a) to (d) for the following questions.

(1) If $X_1 \to F_{(9, 7)}$, $X_2 \to F_{(11, 7)}$ and $X_3 \to F_{(12, 7)}$ then which of the following is true ?

  (a) E (X_1) = E (X_2) = E (X_3) 　　　　(b) E (X_1) < E (X_2) < E (X_3)

  (c) E (X_1) > E (X_2) > E (X_3) 　　　　(d) E (X_2^2) = E (X_1) E (X_3)

(2) If a random variable X follows F-distribution with (6, n) d.f. with mean 2 then the value of n is

  (a) 2 　　　　(b) 3 　　　　(c) 6 　　　　(d) 4

(3) If $X \to F_{(n, n)}$ then mode of the distribution is

  (a) $\dfrac{n + 2}{n - 2}$ 　　　(b) $\dfrac{n - 2}{n + 2}$ 　　　(c) $\dfrac{n (n + 2)}{n (n - 2)}$ 　　　(d) $\dfrac{n (n - 2)}{n (n + 2)}$

(4) If $X \to F_{(8, 8)}$ then mode of the distribution is

  (a) 8 　　　(b) $\dfrac{1}{8}$ 　　　(c) $\dfrac{8}{10}$ 　　　(d) $\dfrac{10}{8}$

(5) Let X follows F-distribution with (10, 11) d.f. and Y = $\dfrac{1}{X}$. Hence mean of the distribution of Y is

  (a) $\dfrac{11}{10}$ 　　　(b) $\dfrac{11}{9}$ 　　　(c) $\dfrac{10}{8}$ 　　　(d) $\dfrac{11}{8}$

(6) Suppose $X_1, X_2 ... X_{10}$ are i.i.d. N (0, 1) variates and

  $U = \dfrac{X_1^2 + X_2^2 + X_3^2 + X_4^2 + X_5^2}{X_6^2 + X_7^2 + X_8^2 + X_9^2 + X_{10}^2}$. Then median of distribution of U is

  (a) 1 　　　　(b) 2 　　　　(c) 5 　　　　(d) 1/2

(7) If $X \to F_{9, 7}$, $Y \to F_{7, 9}$ and P $(X \ge 10)$ + P $(Y \ge K)$ = 1 then the value of constant K is

  (a) 8 　　　　(b) 9 　　　　(c) $\dfrac{9}{10}$ 　　　(d) $\dfrac{1}{10}$

(8) Suppose X_1 and X_2 are independent variables such that $X_1 \to G (\alpha, \lambda_1)$ and

  $X_2 \to G (\alpha, \lambda_2)$ then the probability distribution of $\dfrac{\lambda_2 X_1}{\lambda_1 X_2}$ is

  (a) $F_{(2\lambda_2, 2\lambda_1)}$ 　　　　　　(b) $F_{(2\lambda_1, 2\lambda_2)}$

  (c) $F_{(\lambda_1, \lambda_2)}$ 　　　　　　(d) $F_{(2\lambda_2, 2\lambda_1)}$

(9) Let X and Y be two independent G (16, 4) and G (6, 4) variates respectively then the mean of the distribution of $\dfrac{8X}{3Y}$ is

(a) $\dfrac{8}{3}$ (b) $\dfrac{6}{4}$ (c) $\dfrac{16}{3}$ (d) $\dfrac{8}{6}$

(10) Let X_1 and X_2 be two independent G (16, 4) and G (6, 4) variates respectively. Hence the distribution of $\dfrac{3Y}{8X}$ is

(a) $F_{(8, 8)}$ (b) $F_{(3, 8)}$ (c) $F_{(8, 3)}$ (d) $F_{(6, 16)}$

(11) If X_1 and X_2 are two independent N (0, 1) and $N\left(0, \dfrac{1}{2}\right)$ variates respectively then which of the following will be the distribution of $\dfrac{X_1^2}{2X_2^2}$?

(a) $F_{(1, 2)}$ (b) $F_{(2, 1)}$ (c) $F_{(1, 1)}$ (d) $F_{(2, 2)}$

(12) Let X_1, X_2, X_3 be independent standard normal variates. Then the distribution of $\dfrac{2X_3^2}{X_1^2 + X_2^2}$ is

(a) $F_{(2, 1)}$ (b) $F_{(1, 2)}$ (c) $F_{(2, 2)}$ (d) $F_{(1, 1)}$

(13) Suppose X_1, X_2, ..., X_{10} and Y_1, Y_2, ..., Y_9 are random samples from N (μ_1, σ^2) and N (μ_2, σ^2) respectively $s_1^2 = \dfrac{\sum (X_i - \bar{X})^2}{9}$ and $s_2^2 = \dfrac{\sum (Y_i - \bar{Y})^2}{8}$. Then distribution of $\dfrac{s_1^2}{s_2^2}$ is

(a) $F_{(10, 9)}$ (b) $F_{(9, 10)}$ (c) $F_{(8, 9)}$ (d) $F_{(9, 8)}$

(14) Let X have F-distribution with 4 and 8 d.f. If P [X ≤ b] = 0.01 then value of b is

(a) 14.8   (b) $\dfrac{1}{14.8}$ (c) 7.01   (d) $\dfrac{1}{7.01}$

(15) If mode of F-distribution with 5 and n_2 d.f. is 0.4 then value of n_2 is

(a) 3   (b) 4   (c) 5   (d) 6

(16) If X → F (3, 6) and Y → F (6, 3) then the value of P (X ≥ 10) + P (Y ≥ 0.1) is

(a) 1   (b) 0.1   (c) 0.9   (d) 0.5

(17) If X → $F_{(n_1, n_2)}$ then E (X) is

(a) $\dfrac{n_1}{n_1 - 2}$ (b) $\dfrac{n_2}{n_1 - 2}$ (c) $\dfrac{n_2}{n_2 - 2}$ (d) $\dfrac{n_1}{n_1 + n_2}$

(18) If X → F (4, 4) then the median of X is

(a) 4   (b) 3   (c) 3.5   (d) 1

(19) If X follows F (6, 8) distribution and Y follows F (8, 6) distribution such that P (X ≥ 9) + P (Y ≥ k) = 1 then value of k is

(a) 9   (b) $\dfrac{1}{9}$ (c) 0.9   (d) 0.1

Exercise 3 (C)

State whether the following statements are True or False.

(1) The rth raw moment of F variate with (n, n) d.f. $\mu_r' = \dfrac{B\left(\dfrac{n}{2} + r, \dfrac{n}{2} - r\right)}{B\left(\dfrac{n}{2}, \dfrac{n}{2}\right)}$ if $n > 2r$.

(2) If $X \to F_{n_1, n_2}$ then r^{th} raw moment of the distribution of X is

$$\left(\frac{n_1}{n_2}\right)^r \frac{B\left(\dfrac{n_1}{2} + r, \dfrac{n_2}{2} - r\right)}{B\left(\dfrac{n_1}{2}, \dfrac{n_2}{2}\right)} \text{ if } \frac{n_2}{2} > r.$$

(3) If X follows F-distribution with (n_1, n_2) d.f. then mean of X is independent of n_1.

(4) Suppose X has F-distribution with (2n, 2n) d.f.. Hence median of the distribution is 2.

(5) If Y follows F-distribution with $(7, n_2)$ d.f. then the probability distribution of 7Y tends to t-distribution with 7 d.f. as $n_2 \to \infty$.

(6) If X follows F-distribution with (2, n) d.f. then $\displaystyle \lim_{n \to \infty} P(X \geq 4) = e^{-4}$.

(7) If $X \to F_{(9, 8)}$ and $Y = \dfrac{1}{X}$ then $E(Y) = \dfrac{9}{8}$.

(8) If X has F-distribution with (10, 10) d.f., then $Q_1 Q_3 = 10$ where, Q_1 and Q_3 are first and third quartiles of the distribution respectively.

(9) Let X and Y be F variates with (8, 7) and (7, 8) d.f. respectively, then
$$P(X \geq 8) + P(Y \geq 7) = 1$$

(10) Suppose X_1 and X_2 are F variates with (11, 9) and (9, 11) d.f. respectively. Then
$$P(X_1 \geq a) + P\left(X_2 \geq \frac{1}{a}\right) = 1 \text{ where a is constant.}$$

(11) If $X \to$ t-distribution with 12 d.f. then $X^2 \to F_{12, 1}$.

(12) Suppose $X_1, X_2, ..., X_{13}$ is a random sample from $N(\mu_1, \sigma^2)$ distribution and $Y_1, Y_2, ..., Y_{11}$ is random sample from $N(\mu_2, \sigma^2)$. Let $s_1^2 = \dfrac{\sum_{i=1}^{13} (X_i - \bar{X})^2}{12}$ and $s_2^2 = \dfrac{\sum_{i=1}^{11} (Y_i - \bar{Y})^2}{10}$. Then $\dfrac{s_1^2}{s_2^2}$ has F-distribution 12 and 10 d.f.

(13) Mean of $F_{4, 10}$ distribution is same as mean of $F_{6, 10}$ distribution.

(14) If $X \to F_{(n_1, n_2)}$ then $1/X \sim> F_{(n_2, n_1)}$.

(15) F distribution with parameter n, n is symmetric.

(16) Mean of F-distribution with (n_1, n_2) d.f. greater than mode.

(17) If X_1, X_2, X_3, X_4 are i.i.d. N (0, 1) variates then $\dfrac{3X_4^2}{X_1^2 + X_2^2 + X_3^2}$ has F-distribution with $F_{3,3}$ d.f.

Answers of Exercise 3 (A)

(9) $f(u) = 1/(1 + u)^2$, $u > 0$

(10) (i) $F_{(n_2, n_1)}$ (ii) $f(u) = \dfrac{1}{B\left(\dfrac{n_1}{2}, \dfrac{n_2}{2}\right)} u^{\frac{n_2}{2} - 1} (1 - u)^{\frac{n_2}{2} - 1}$, $0 < u < 1$

(12) $f(u) = \dfrac{2^{n-3}}{B\left(\dfrac{1}{2}, \dfrac{n-2}{2}\right)} \cdot \dfrac{Y^{\frac{n-4}{2}}}{(1 + y)^{n-2}}$, $y > 0$

Using duplication formula it can be written as p.d.f. of F (n – 2, n – 2) with d.f.

(15) 1

(16) $a = 0.01006$, $b = 8.65$ (17) $k = \dfrac{(n_1/n_2)^{\frac{n_1}{2}}}{B\left(\dfrac{n_1}{2}, \dfrac{n_2}{2}\right)}$, $F_{(n_2, n_1)}$.

(18) Answer is not unique, one of the answer is $\dfrac{2\,(X_1 - 1)^2}{\left(\dfrac{X_2 - 2}{2}\right)^2 + \left(\dfrac{X_3 - 3}{3}\right)^2}$

(19) 1/5403   (21) 21.2   (23) $a = \dfrac{1}{5982}$; $b = 11.26$

Answers of Exercise 3 (B)

(1) a	(2) d	(3) b	(4) c
(5) c	(6) a	(7) d	(8) b
(9) d	(10) a	(11) c	(12) b
(13) d	(14) b	(15) b	(16) a
(17) c	(18) d	(19) b	

Answers of Exercise 3 (C)

(1) True	(2) False	(3) True	(4) False
(5) False	(6) True	(7) False	(8) False
(9) False	(10) True	(11) False	(12) True
(13) True	(14) True	(15) True	(16) True
(17) False.			

Chapter 4...

Sampling Distributions

Friedrich Robert Helmert

Friedrich Robert Helmert (July 31, 1843 – June 15, 1917) was a German geodesistand an important writer on the theory of errors. Helmert was born in Freiberg, Kingdom of Saxony. After schooling in Freiberg and Dresden, he entered the Polytechnische Schule, now Technische Universität, in Dresden to study engineering science in 1859. Finding him especially enthusiastic about geodesy, one of his teachers, August Nagel, hired him while still a student to work on the triangulation of the Erzgebirge and the drafting of the trigonometric network for Saxony. In 1863 Helmert became Nagel's assistant on the Central European Arc Measurement. After a year's study of mathematics and astronomy Helmert obtained his doctor's degree from the University of Leipzig in 1867 for a thesis based on his work for Nagel..

Contents ...

Key Words :

Random sample, Parameter, Statistic, Sampling distribution of statistic, Standard Error (S.E.), Orthogonal transformation.

Objectives :

(1) To introduce the concepts of random sample, parameter, statistic, Standard Error (S.E.) to students used in testing of hypothesis and statistical inference.

(2) To prove the result mean and variance of sample drawn from N (μ, σ^2) are independent. It is rather surprising.

(3) To understand that $\dfrac{n\,s^2}{\sigma^2} \to \chi^2_{n-1}$.

(4) To apply the above two results to solve the problems on computing probabilities.

4.0 Introduction

In order to draw inference about a certain phenomenon, sampling is a well accepted tool. Entire population cannot be studied due to several reasons. In such a situation sampling is the only alternative. A properly drawn sample is much useful in drawing reliable conclusions. Here, we draw a sample from probability distribution rather than a group of objects. Using simulation technique sample is drawn.

4.1 Random Sample from a Continuous Distribution

A random sample from a continuous probability distribution $f(x, \theta)$ is nothing but the values of independent and identically distributed random variables with the common probability density function $f(x, \theta)$.

Definition : Random sample : If $X_1, X_2, \ldots, X_n$ are independent and identically distributed random variables, with p.d.f. $f(x, \theta)$, then we say that, they form a random sample from the population with p.d.f. $f(x, \theta)$.

Note :

(1) For drawing inference, we use the numerical values of $X_1, X_2, \ldots, X_n$.

(2) The joint p.d.f. of $X_1, X_2, \ldots X_n$ is,

$$f(x_1, x_2, \ldots, x_n) = f(x_1), f(x_2), \ldots, f(x_n) = \prod_{i=1}^{n} f(x_i)$$

4.2 Statistic and Parameter

Using the random sample $X_1, X_2, \ldots, X_n$ we draw conclusion about the unknown probability distribution. However probability distribution can be studied if the parameter θ is known. In other words study of probability distribution reduces to the study of parameter θ. We use sampled observations for this purpose. There are various ways of summarizing the sampled observations. The summarized quantity is called as statistic. We define it precisely as follows.

Definition : If $X_1, X_2, \ldots, X_n$ is a random sample from a probability distribution $f(x, \theta)$, then $T = T(x_1, x_2, \ldots, x_n)$ a function of sample values which does not involve unknown parameter θ is called as a *statistic (or estimator)*.

Some typical statistics are given below :

(i)   Sample mean : $T = T(x_1, x_2, \ldots, x_n) = \dfrac{\sum x_i}{n}$

$\therefore$ $T = \bar{X}$ is a statistic

(ii) Sample variance :

$$T = T(x_1, x_2, \ldots x_n)$$

$$= \frac{1}{n-1} \sum (x_i - \bar{x})^2 \text{ is a statistic.}$$

(iii) $T = \max\{x_1, x_2, \ldots, x_n\}$, $T = \min\{x_1, x_2, \ldots, x_n\}$

$T = $ median of $\{x_1, x_2, \ldots, x_n\}$ are also statistics.

(iv) Sample proportion (p) of observations less than μ is also a statistic.

Let, $Y_i = 1$; if $X_i < \mu$

$= 0$; if $X_i \geq \mu$

then, $T = T(x_1, x_2, \ldots, x_n)$

$$= \phi(y_1, y_2, \ldots, y_n) = \frac{\sum y_i}{n}$$

$$= \frac{\text{No. of observations less than } \mu}{n} = p$$

$\therefore$ p is a statistic.

In this manner several statistics are defined and the statistic suitable for the purpose is selected.

Note :

(1) Verbally, statistic is a summarized quantity of sample values such as mean, variance, proportion, correlation coefficient. On the other hand similar quantities which correspond to population are called as **parameters**. If $f(x, \theta)$ is a p.d.f. then the constant θ involved in it is also called as **parameter**. Note that, the population mean, variance etc. are the functions of θ. Thus by the term parameter we mean either θ or function of θ.

(2) A noteworthy difference in statistic and parameter is that the former is a random variable and the latter is a constant.

Since statistics is a function of random variables $X_1, X_2, \ldots, X_n$ it is also a random variable. It varies from sample to sample.

(3) Since statistic is a random variable it possesses some probability distribution, it may not be the same as that of the parent distribution $f(x, \theta)$. However, parameter being a constant, does not possess probability distribution.

4.3 Sampling Distribution of Statistic and Standard Error

Further statistical inference is based on statistic, therefore we need to study its probability distribution. The general theory is developed in the subsequent discussion.

Definition : If $X_1, X_2, \ldots, X_n$ is a random sample from $f(x, \theta)$ then the probability distribution of statistic $T(x_1, x_2, \ldots, x_n)$ is called as its *sampling distribution* and standard deviation of T is called as its *standard error (S.E.)*.

$\therefore$ S.E. $(T) = \sqrt{E[T - E(T)]^2}$

In the problem of testing of hypothesis standard error of statistic used in the test procedure turns out to be important quantity. We list below some typical statistics alongwith standard errors.

(i) $T = \bar{X}$,   S.E. $(T) = \dfrac{\sigma}{\sqrt{n}}$

  where, σ^2 is the population variance.

(ii)   Suppose there are two samples of size n_1 and n_2 respectively. First is drawn from a population with variance σ_1^2 and the second is drawn from a population with variance σ_2^2.

Let, $T = \bar{X}_1 - \bar{X}_2$, then

$$\text{S.E. } (T) = \sqrt{\text{Var } (\bar{X}_1 - \bar{X}_2)} = \sqrt{\frac{\sigma_1^2}{n_1} + \frac{\sigma_2^2}{n_2}}$$

(iii) If p is the sample proportion then

$$\text{S.E. } (p) = \sqrt{\frac{PQ}{n}}$$

where, P : population proportion and $Q = 1 - P$.

(iv) If p_1 and p_2 are proportions obtained using two samples one from first population with population proportion P_1 and the other sample from second population with P_2 as population proportion. Let, $T = P_1 - P_2$ then,

$$\text{S.E. } (T) = \sqrt{\frac{P_1 Q_1}{n_1} + \frac{P_2 Q_2}{n_2}}$$

(v)   If $T = S^2 = \dfrac{\Sigma (X_i - \bar{X})^2}{n}$

then,   S.E. $(T) = \sigma^2 \sqrt{\dfrac{2(n-1)}{n^2}} \approx \sigma^2 \sqrt{\dfrac{2}{n}}$.

(vi) If T is sample median of a sample from N (μ, σ^2), then

$$\text{S.E. } (T) = 1.2533 \frac{\sigma}{\sqrt{n}}.$$

4.4 Sampling Distribution of Mean and Variance from Normal Population

In majority cases a sample is assumed to be drawn from normal population. In order to develop tests regarding mean μ and variance σ^2 we need to find sampling distribution of mean $\bar{X}$ and sampling variance $S^2 = \dfrac{1}{n} \Sigma (X_i - \bar{X})^2$. The derivation of distribution of S^2 needs orthogonal transformation and some results. Those are introduced and included in brief in the following discussion.

Transformation of Variables :

Suppose $X_1, X_2, ..., X_n$ is a random sample, we define new set of variables $Y_1, Y_2, ... , Y_n$ using linear transformation as follows :

$$Y_1 = a_{11} X_1 + a_{12} X_2 + ... + a_{1n} X_n$$

$$Y_2 = a_{21} X_1 + a_{22} X_2 + ... + a_{2n} X_n$$

$$..$$

$$..$$

$$Y_n = a_{n1}X_1 + a_{n2}X_2 + ... + a_{nn} X_n$$

The above transformation can be written in matrix notation as $Y = AX$ where,

$$Y = \begin{bmatrix} Y_1 \\ Y_2 \\ \vdots \\ Y_n \end{bmatrix}, \quad A = \begin{bmatrix} a_{11} & a_{12} & \cdots & a_{1n} \\ a_{21} & a_{22} & \cdots & a_{2n} \\ & & & \\ a_{n1} & a_{n2} & \cdots & a_{nn} \end{bmatrix} \quad \text{and} \quad X = \begin{bmatrix} X_1 \\ X_2 \\ \vdots \\ X_n \end{bmatrix}$$

Orthogonal Transformation :

Matrix transformation $Y = AX$ is said to be orthogonal if the matrix A is orthogonal.

Note :

(1) A is orthogonal matrix if $A'A = AA' = I$, in other words $A^{-1} = A'$.

(2) If A is orthogonal matrix then $|A| = 1$ or -1.

(3) If the first row of orthogonal matrix is decided the remaining rows can be selected suitably.

Illustrations (1) : $A = \begin{bmatrix} 1/\sqrt{2} & 1/\sqrt{2} \\ 1/\sqrt{2} & -1/\sqrt{2} \end{bmatrix}$ is orthogonal matrix hence

$$Y_1 = \frac{X_1}{\sqrt{2}} + \frac{X_2}{\sqrt{2}} \text{ and } Y_2 = \frac{X_1}{\sqrt{2}} - \frac{X_2}{\sqrt{2}} \text{ is an orthogonal transformation.}$$

(2) $A = \begin{bmatrix} 1/3 & 2/3 & 2/3 \\ 2/3 & 1/3 & -2/3 \\ -2/3 & 2/3 & -1/3 \end{bmatrix}$ is an orthogonal matrix. Hence the corresponding

transformation $Y_1 = \dfrac{1}{3} (X_1 + 2 X_2 + 2 X_3)$, $Y_2 = \dfrac{1}{3} (2X_1 + X_2 - 2 X_3)$ and

$$Y_3 = \frac{1}{3} (-2 X_1 + 2X_2 - X_3) \text{ is orthogonal.}$$

Result : If $Y = AX$ is an orthogonal transformation then,

(a) Jacobian of transformation

$$|J| = \left| \frac{\partial (X_1, X_2, ..., X_n)}{\partial (Y_1, Y_2, ..., Y_n)} \right| = 1$$

(b) $\qquad \sum Y_i^2 = \sum X_i^2$

Proof : (a) Note that,

$$\frac{\partial (X_1, X_2,..., X_n)}{\partial (Y_1, Y_2, ..., Y_n)} = \frac{1}{\dfrac{\partial (Y_1, Y_2, ..., Y_n)}{\partial (X_1, X_2,..., X_n)}} = \frac{1}{J}$$

$$\therefore \frac{1}{J} = \begin{bmatrix} \dfrac{\partial y_1}{\partial x_1} & \dfrac{\partial y_1}{\partial x_2} & \cdots\cdots & \dfrac{\partial y_1}{\partial x_n} \\[2mm] \dfrac{\partial y_2}{\partial x_1} & \dfrac{\partial y_2}{\partial x_2} & \cdots\cdots & \dfrac{\partial y_2}{\partial x_n} \\[2mm] \cdots\cdots\cdots\cdots\cdots\cdots \\[2mm] \dfrac{\partial y_n}{\partial x_1} & \dfrac{\partial y_n}{\partial x_2} & \cdots\cdots & \dfrac{\partial y_n}{\partial x_n} \end{bmatrix} = \begin{bmatrix} a_{11} & a_{12} & \cdots & a_{1n} \\ a_{21} & a_{22} & \cdots & a_{2n} \\ \cdots\cdots\cdots\cdots\cdots \\ a_{n1} & a_{n2} & \cdots & a_{nn} \end{bmatrix} = A$$

$$\left|\frac{1}{J}\right| = |A| = 1 \text{ or} -1 \qquad \therefore \ |J| = 1$$

(b) Note that, $\sum X_i^2 = [x_1, x_2, ..., x_n] \begin{bmatrix} x_1 \\ x_2 \\ ... \\ x_n \end{bmatrix} = X' X$

Similarly, $\qquad \sum Y_i^2 = Y' Y = (AX)' (AX)$

$$= X'A'AX \qquad\qquad (\because A'A = I)$$

$$= X' I X = X' X = \sum X_i^2$$

Theorem 1 :

If $X_1, X_2, ..., X_n$ is a random sample from $N(\mu, \sigma^2)$ then,

(i) Sample mean $\bar{X}$ and sample variance $S^2 = \dfrac{1}{n} \sum (X_i - \bar{X})^2$ are independently distributed.

(ii) $\bar{X}$ follows $N(\mu, \sigma^2/n)$

(iii) $\dfrac{nS^2}{\sigma^2} = \dfrac{\sum (X_i - \bar{X})^2}{\sigma^2}$ follows chi-square distribution with $(n-1)$ degrees of freedom.

Proof : Let $Y = AX$ be an orthogonal transformation (or Helmert transformation) with $Y_1 = \dfrac{1}{\sqrt{n}} (X_1 + X_2 + ... + X_n)$ and $Y_2, Y_3, ..., Y_n$ be suitably defined.

Clearly $Y_1 = \dfrac{1}{\sqrt{n}} \sum X_i = \dfrac{n\bar{X}}{\sqrt{n}} = \sqrt{n}\,\bar{X}$ $\qquad\qquad$ ... (1)

$\therefore$　Joint p.d.f. of $(Y_1, Y_2, ..., Y_n)$ = Joint p.d.f. of $(X_1, X_2, ... , X_n)\ |J|$.

Since,　　$|J| = \left| \dfrac{\partial\,(x_1, x_2, ..., x_n)}{\partial\,(y_1, y_2, ...,y_n)} \right| = 1$

$p\,(y_1, y_2, ..., y_n) = f\,(x_1) \cdot f\,(x_2) .... f\,(x_n) \cdot 1$

$$= \left(\frac{1}{\sigma\sqrt{2\pi}} \right)^n e^{-\frac{1}{2\sigma^2}\Sigma\,(x_i - \mu)^2} \qquad \text{... (2)}$$

We express $\Sigma\,(x_i - \mu)^2$ in terms of $y_1, y_2, ... y_n$.

Note that, $\Sigma\,(x_i - \mu)^2 = \Sigma\,[(x_i - \bar{x}) + (\bar{x} - \mu)]^2$

$$= \sum_{1}^{n} (x_i - \bar{x})^2 + n\,(\bar{x} - \mu)^2 + 2\,(\bar{x} - \mu)\,\Sigma\,(x_i - \bar{x})$$

$$= \sum_{1}^{n} x_i^2 - n\,\bar{x}^2 + (\sqrt{n}\,\bar{x} - \sqrt{n}\,\mu)^2 + 0 \qquad \text{... from (1)}$$

$$= \sum_{1}^{n} y_i^2 - y_1^2 + (y_1 - \sqrt{n}\,\mu)^2$$

$$= \sum_{2}^{n} y_i^2 + (y_1 - \sqrt{n}\,\mu)^2 \qquad \text{... (3)}$$

Using equation (3) the equation (2) can be written as,

$$p\,(y_1, y_2, ..., y_n) = \left(\frac{1}{\sigma\sqrt{2\pi}} \right)^n e^{-\frac{1}{2\sigma^2}[(y_1 - \sqrt{n}\,\mu)^2 + \sum_{2}^{n} y_i^2]}$$

$$= \left(\frac{1}{\sigma\sqrt{2\pi}} \right) e^{-\frac{1}{2\sigma^2}(y_1 - \sqrt{n}\,\mu)^2} \times \prod_{i=2}^{n} \frac{1}{\sigma\sqrt{2\pi}}\, e^{-\frac{1}{2\sigma^2}y_i^2} \qquad \text{... (4)}$$

From equation (4) we get the following findings

(a)　$Y_1, Y_2, ... , Y_n$ are independently distributed random variables.

(b)　Y_1 follows $N\,(\sqrt{n}\,\mu, \sigma^2)$.

(c)　$Y_2, Y_3, ... , Y_n$ are i.i.d. $N\,(0, \sigma^2)$

(i)　Clearly $\bar{X} = \dfrac{Y_1}{\sqrt{n}}$ and

$$S^2 = \frac{1}{n}\,\Sigma\,(X_i - \bar{X})^2 = \frac{1}{n}\,[\Sigma\,X_i^2 - n\bar{X}^2]$$

$\therefore$

$$S^2 = \frac{1}{n}\left[\sum_{1}^{n} y_i^2 - y_1^2 \right] = \frac{1}{n}\sum_{2}^{n} y_i^2 \qquad \text{... (5)}$$

$\therefore$ $\bar{X}$ is a function of Y_1 alone and S^2 is a function of $(Y_2, Y_3, ..., Y_n)$. Since Y_i's are independent random variables, $\bar{X}$ and S^2 are also independent random variables.

(ii)   Note that, $\bar{X} \to N(\mu, \sigma^2/n)$

(iii)   Since, $Y_i \to N(0, \sigma^2)$ $i = 2, 3, ..., n$

$\therefore$ $\dfrac{Y_i}{\sigma} \to N(0, 1)$

$\therefore$ $\dfrac{Y_i^2}{\sigma^2} \to \chi^2$ with 1 degree of freedom.

Moreover Y_i's are independent random variables, therefore by using additive properties of chi-square we get,

$$\sum_{2}^{n} Y_i^2 /\sigma^2 \to \chi^2 \text{ with } (n-1) \text{ degrees of freedom.}$$

Note that $\displaystyle\sum_{2}^{n} Y_i^2/\sigma^2 = \dfrac{1}{\sigma^2}\sum_{2}^{n} Y_i^2 = \dfrac{1}{\sigma^2} nS^2$ from (5)

We get, $\dfrac{nS^2}{\sigma^2} \to \chi_{n-1}^2$.

Note : It is surprising that $\bar{X}$ and $S^2 = \dfrac{1}{n}\sum(X_i - \bar{X})^2$ both are functions of $X_1, X_2, ..., X_n$ however they are independent random variables. The above theorem is useful and serves good purpose in small sample tests regarding μ and σ^2.

Illustration 2 : If $X_1, X_2, ..., X_{25}$ is a random sample from $N(5, 9)$, then find $P(3.8 \le \bar{X} \le 5.6, 5.637 \le S^2 \le 11.951)$.

Solution : Note that $\bar{X}$ and S^2 are independent random variables, hence

$P(3.8 \le \bar{X} \le 5.6, 5.637 \le S^2 \le 11.951)$

$= P(3.8 \le \bar{X} \le 5.6) \cdot P(5.637 \le S^2 \le 11.951)$

$= P\left(\dfrac{3.8 - 5}{3/5} \le \dfrac{\bar{X} - \mu}{\sigma/\sqrt{n}} \le \dfrac{5.6 - 5}{3/5}\right) \cdot P\left(\dfrac{25}{9} 5.637 \le \dfrac{nS^2}{\sigma^2} \le \dfrac{25}{9} \cdot 11.751\right)$

$= P(-2 \le N(0, 1) \le 1) \cdot P(15.658 \le \chi_{24}^2 \le 33.197)$

$= (1 - 0.15866 - 0.02275) \cdot [P(\chi_{24}^2 > 15.658) - P(\chi_{24}^2 > 33.197)]$

$= 0.81859 \times (0.9 - 0.1) = 0.654872$

Illustration 3 : Let X be a random sample of size 16 be drawn from a normal distribution with $\mu = 5$ and variance $\sigma^2 = 64$. Calculate $P(0 < \bar{X} < 10, 20.92 < S^2 < 122.32)$; $\bar{X}$ and S^2 being respectively the mean and variance of sample.   **(P.U. Oct. 2006)**

Solution : Note that $\bar{X}$ and S^2 are independent random variables.

Also $\dfrac{\bar{X} - \mu}{\sigma/\sqrt{n}} = \dfrac{\bar{X} - 5}{2} \to N(0, 1)$ and $\dfrac{nS^2}{\sigma^2} = \dfrac{16S^2}{64} = \dfrac{S^2}{4} \to \chi^2_{15}$.

$\therefore$ $P(0 < \bar{X} < 10,\ 20.92 < S^2 < 122.32)$

$$= P\left[\dfrac{0-5}{2} < \dfrac{\bar{X}-5}{2} < \dfrac{10-5}{2}\right] P\left[\dfrac{20.92}{4} < \dfrac{S^2}{4} < \dfrac{122.32}{4}\right]$$

$$= P\left[-2.5 < N(0, 1) < 2.5\right] P\left[5.23 < \chi^2_{15} < 30.58\right]$$

$$= \left[1 - P(N(0, 1) > 2.5\right] P\left\{[\chi^2_{15} \geq 5.23] - P[\chi^2_{15} \geq 5.23] - P[\chi^2_{15} \geq 30.58]\right\}$$

$$= [1 - 2(0.0062097)] [0.99 - 0.01]$$

$$= (0.9875806)(0.98) = \mathbf{0.967829}$$

Illustration 4 : Let $\bar{X}$ and S^2 be the mean and variance of a random sample of size 25 from N (3, 100) distribution.

Evaluate $P[0 < \bar{X} < 6,\ 55.2 < S^2 < 145.6]$

Solution : $P[0 < \bar{X} < 6,\ 55.2 < S^2 < 145.6]$

$$= P[0 < \bar{X} < 6] P[55.2 < S^2 < 145.6] \quad (\because \bar{X} \text{ and } S^2 \text{ are independent})$$

$$= P\left[\dfrac{0-3}{2} < \dfrac{\bar{X}-3}{2} < \dfrac{6-3}{2}\right] P\left[\dfrac{55.2}{4} < \dfrac{S^2}{4} < \dfrac{145.6}{4}\right]$$

$$= P\left[-\dfrac{3}{2} < N(0, 1) < \dfrac{3}{2}\right] P[13.8 < \chi^{224} < 36.4]$$

$$= [1 - 2(0.06807)] [P(\chi^2_{24} \geq 13.8) - P(\chi^2_{24} \geq 36.4)]$$

$$= [0.863859][0.99 - 0.05] = \mathbf{0.812027}$$

Note : $\dfrac{\bar{X}-3}{10/5} = \dfrac{\bar{X}-3}{2} \to N(0, 1)$ and $\dfrac{nS^2}{\sigma^2} = \dfrac{25S^2}{100} = \dfrac{S^2}{4}$

Solved Examples

Example 4.1 : If $X_1, X_2, ..., X_n$ is a random sample from N (μ, σ^2) then probability

distribution of $\dfrac{(\bar{X} - \mu)\sqrt{n}}{\sqrt{\dfrac{1}{n-1}\Sigma(X_i - \bar{X})^2}}$ is t with $(n - 1)$ d.f.

Solution : We know that $\bar{X} \to N(\mu, \sigma^2/n)$ and $\dfrac{\Sigma(X_i - \bar{X})^2}{\sigma^2} \to \chi^2$ with $(n - 1)$ d.f.,

Hence $\dfrac{\bar{X} - \mu}{\sigma/\sqrt{n}} \to N(0, 1)$. Since $\bar{X}$ and S^2 are independent we get,

$$W = \frac{\text{Standard Normal Variate}}{\sqrt{\text{Chi-square variate/d.f.}}} = \frac{(\bar{X} - \mu) \div \sigma/\sqrt{n}}{\sqrt{\dfrac{\sum (X_i - \bar{X})^2}{\sigma^2} \div (n - 1)}}$$

$$= \frac{(\bar{X} - \mu) \sqrt{n}}{\sqrt{\dfrac{1}{n - 1} \sum (X_i - \bar{X})^2}} \to t \text{ with } (n - 1) \text{ d.f.}$$

Example 4.2 : If $X_1, X_2, \ldots, X_{n_1}$ is a random sample from $N(\mu_1, \sigma^2)$ and $Y_1, Y_2, \ldots, Y_{n_2}$ is a random sample from $N(\mu_2, \sigma^2)$ then with usual notation show that

(a) $\dfrac{(\bar{X} - \bar{Y}) - (\mu_1 - \mu_2)}{\sqrt{S^2 \left(\dfrac{1}{n_1} + \dfrac{1}{n_2}\right)}} \to$ Students t-distribution with $n_1 + n_2 - 2$ d.f.

where, $S^2 = \dfrac{1}{n_1 + n_2 - 2} \left[\sum (X_i - \bar{X})^2 + \sum (Y_i - \bar{Y})^2\right]$

(b) $\dfrac{\dfrac{\sum (X_i - \bar{X})^2}{n_1 - 1}}{\dfrac{\sum (Y_i - \bar{Y})^2}{n_2 - 1}} \to$ F-distribution with parameters $n_1 - 1$ and $n_2 - 1$.

Solution : Note that $\bar{X}$ and $S_1^2 = \dfrac{1}{n_1} \sum (X - \bar{X})^2$ are independent as well as $\bar{Y}$ and $S_2^2 = \dfrac{1}{n_2} \sum (Y - \bar{Y})^2$ are also independent. Moreover since $\bar{X}, S_1^2$ and $\bar{Y}, S_2^2$ are based on independent samples $\bar{X}, \bar{Y}, S_1^2$ and S_2^2 are independent random variables.

$\therefore \quad \bar{X} - \bar{Y} \to N\left(\mu_1 - \mu_2, \sigma^2/n_1 + \sigma^2/n_2\right)$

$\therefore \quad U = \dfrac{(\bar{X} - \bar{Y}) - (\mu_1 - \mu_2)}{\sqrt{\dfrac{\sigma^2}{n_1} + \dfrac{\sigma^2}{n_2}}} \to N(0, 1)$

Since $\dfrac{\sum (X_i - \bar{X})^2}{\sigma^2}$ and $\dfrac{\sum (Y_i - \bar{Y})^2}{\sigma^2}$ are independent chi-square variates with $n_1 - 1$ and $n_2 - 1$ d.f. respectively, by additive property we get,

$$V = \frac{\sum (X_i - \bar{X})^2}{\sigma^2} + \frac{\sum (Y_i - \bar{Y})^2}{\sigma^2} \to \chi^2 \text{ with } (n_1 + n_2 - 2) \text{ d.f.}$$

$$\therefore \quad W = \frac{U}{\sqrt{V/(n_1 + n_2 - 2)}} \to t \text{ with } (n_1 + n_2 - 2) \text{ d.f.}$$

$$\therefore \quad W = \frac{(\bar{X} - \bar{Y}) - (\mu_1 - \mu_2)}{\sqrt{\sigma^2 \left(\dfrac{1}{n_1} + \dfrac{1}{n_2} \right)}} \times \frac{\sqrt{n_1 + n_2 - 2}}{\sqrt{\dfrac{1}{\sigma^2} [\sum (X_i - \bar{X})^2 + \sum (Y_i - \bar{Y})^2]}}$$

$$= \frac{(\bar{X} - \bar{Y}) - (\mu_1 - \mu_2)}{\sqrt{S^2 \left(\dfrac{1}{n_1} + \dfrac{1}{n_2} \right)}} \to t_{n_1 + n_2 - 2}$$

(b) Note that $\dfrac{n_1 S_1^2}{\sigma^2} \to \chi^2$ with $(n_1 - 1)$ d.f.

and $\dfrac{n_2 S_2^2}{\sigma^2} \to \chi^2$ with $(n_2 - 1)$ d.f.

Since S_1^2 and S_2^2 are independent

$$F = \frac{\dfrac{n_1 S_1^2}{\sigma^2} \div (n_1 - 1)}{\dfrac{n_2 S_2^2}{\sigma^2} \div (n_2 - 1)} \to \text{F-distribution with d.f. } (n_1 - 1) \text{ and } (n_2 - 1).$$

$$\therefore \quad F = \frac{\dfrac{\sum (X_i - \bar{X})^2}{n_1 - 1}}{\dfrac{\sum (Y_i - \bar{Y})^2}{n_2 - 1}} \to F_{n_1-1,\, n_2-1}$$

Points to Remember

1. If $X_1, X_2, \ldots, X_n$ is a random sample from a probability distribution then $T = T(x_1, x_2, \ldots, x_n)$ a function of sample values which does not involve unknown parameter is called a statistic. e.g.

(i) $T = $ sample mean $= \dfrac{\sum\limits_{i=1}^{n} X_i}{n}$.

(ii) $T = \dfrac{1}{n-1} \Sigma (X_i - \bar{X})^2$.   (iii) T = sample proportion = p.

(iv) $T = \max \{X_1, X_2, ..., X_n\}$ or $T = \min \{X_1, X_2, ..., X_n\}$.

2. Statistic is a variable while parameter is a constant.

3. If T is a statistic then Standard Error (S.E.) of T is $\text{S.E. } (T) = \sqrt{E\,[T - E(T)]^2}$.

4.

Statistic	Standard Error (S.E.)
$T = \bar{X}$ = Sample mean.	$\text{S.E. } (T) = \dfrac{\sigma}{\sqrt{n}}$ where, σ^2 is population variance and n is sample size.
$T = \bar{X}_1 - \bar{X}_2$ where, $\bar{X}_1$ and $\bar{X}_2$ are means of samples of sizes n_1, n_2 drawn from 1st and 2nd population respectively.	$\text{S.E. } (T) = \sqrt{\dfrac{\sigma_1^2}{n_1} + \dfrac{\sigma_2^2}{n_2}}$ where, σ_1^2 and σ_2^2 are population variances and n_1, n_2 are sample sizes.
$T = P$ = Sample proportion.	$\text{S.E. } (P) = \sqrt{\dfrac{P\,(1-P)}{n}}$ where, P : Population proportion.
$T = P_1 - P_2$ where, P_1 and P_2 are proportions in the samples of sizes n_1 and n_2 drawn from.	$\text{S.E. } (T) = \text{S.E. } (P_1 - P_2)$ $= \sqrt{\dfrac{P_1\,(1-P_1)}{n_1} + \dfrac{P_2\,(1-P_2)}{n_2}}$

5. If $X_1, X_2, ..., X_i ... X_n$ is a random sample from $N\,(\mu, \sigma^2)$ then

(i) Sample mean $\bar{X}$ and sample variance $S^2 = \dfrac{1}{n} \sum_{i=1}^{n} (X_i - \bar{X})^2$ are independently distributed.

(ii) $\bar{X}$ follows $N\left(\mu, \dfrac{\sigma^2}{n}\right)$.

(iii) $\dfrac{n\,S^2}{\sigma^2} = \dfrac{\sum_{i=1}^{n} (X_i - \bar{X})^2}{\sigma^2}$ follows chi-square distribution with $(n-1)$ degrees of freedom.

6. If $X_1, X_2, ..., X_n$ is a random sample from $N\,(\mu, \sigma^2)$ then $\dfrac{\bar{X} - \mu}{S/\sqrt{n}}$ where, $S^2 = \dfrac{1}{n-1} (X_i - \bar{X})^2$ follows t-distribution with $(n-1)$ d.f.

7. If $X_1, X_2, ..., X_{n_1}$ is a random sample from $N(\mu_1, \sigma^2)$ and $Y_1, Y_2, ..., Y_{n_2}$ is a random sample from $N(\mu_2, \sigma^2)$ then

(i) $\dfrac{\bar{X} - \bar{Y} - (\mu_1 - \mu_2)}{S\sqrt{\dfrac{1}{n_1} + \dfrac{1}{n_2}}}$ follows t-distribution with $n_1 + n_2 - 2$ d.f where,

$$S^2 = \frac{[\Sigma (X_i - \bar{X})^2 + \Sigma (Y_i - \bar{Y})^2]}{n_1 + n_2 - 2}$$

(ii) $\dfrac{S_1^2}{S_2^2}$ where, $S_1^2 = \dfrac{\Sigma (X_i - \bar{X})^2}{n_1 - 1}$ and $S_2^2 = \dfrac{\Sigma (Y_i - \bar{Y})^2}{n_2 - 1}$ follows F-distribution with $(n_1 - 1)$ and $(n_2 - 1)$ d.f.

Exercise 4 (A)

1. Explain the term random sample from a probability distribution.
2. Explain the following terms :
   (i) Statistic   (ii) Sampling distribution of a statistic   **(P.U. April 2004)**
   (iii) Standard error of a statistic.   **(P.U. Nov. 2004)**
3. Distinguish between 'parameter' and 'statistic'.
4. Obtain the sampling distribution of a mean of a random sample drawn from
       (i) Normal distribution   **(P.U. Oct. 2011)**
       (ii) Exponential distribution
       (iii) Gamma distribution.   **(P.U. Oct. 2012, 1999)**

5. Let $X_1, X_2, ..., X_n$ be a random sample from $N(\mu, \sigma^2)$ then show that $\bar{X}$ and $S^2 = \dfrac{1}{n}(X_i - \bar{X})^2$ are independently distributed random variables. Also find the probability distribution of nS^2/σ^2. Discuss the importance of the result.
6. Explain the term sampling distribution of a statistic. Also obtain the sampling distribution of a mean of random sample of size n drawn from exponential distribution with parameter α.   **(P.U. May 2011)**

Exercise 4 (B)

1. If $X_1, X_2, ..., X_{20}$ is a random sample from $N(\mu, \sigma^2)$, find the probability distribution of

$$\frac{\dfrac{\sum\limits_{i=1}^{10} (X_i - a)^2}{20}}{\sum\limits_{i=11}^{} (X_i - b)^2} \quad \text{where, } a = \frac{1}{10}\sum\limits_{1}^{10} X_i \text{ and } b = \frac{\sum\limits_{i=11}^{20} X_i}{10}.$$

2. (a) If $\bar{X}$ and S^2 are the mean and the variance of a random sample of size 16 from

   N (3, 64), then find P $(-1 < \bar{X} < 5, 34.188 < S^2 < 77.244)$. **(P.U. 1998)**

   (b) Let X_1, X_2, ..., X_{16} be a random sample of size 16 from N (20, 25) population. Find

   P $(19 < \bar{X} < 21.5, 13.354 < S^2 < 34.854)$. **(P.U. Oct. 2005)**

3. If $\bar{X}$ and S^2 denote the mean and variance of a random sample of size 10 from N (4, 160) then evaluate,

   P $[0 < \bar{X} < 4, \ 86.08 < S^2 < 170.496]$ **(P.U. 1997)**

4. Let X_1, X_2, ..., X_{15} be a random sample from a N (3, 10). Determine

   $$P\left[\sum_{i=1}^{15} (X_i - \bar{X})^2 \le 77.9\right].$$ **(P.U. 1998)**

5. Let S^2 be a sample variance of a random sample from N (μ, σ^2), find the E (S^2) and Var (S^2).

6. If X_1, X_2 ... X_{10} is a r.s. from N (8, 16), find E (S^2), V (S^2), P $(S^2 \ge 27.0704)$.

7. Let X_1 and X_2 are i.i.d. N $(0, \sigma^2)$. If $Y_1 = \dfrac{X_1}{\sqrt{2}} + \dfrac{X_2}{\sqrt{2}}$ and $Y_2 = \dfrac{X_1}{\sqrt{2}} - \dfrac{X_2}{\sqrt{2}}$, show that Y_1

   and Y_2 are independent. Hence or otherwise show that $\dfrac{X_1 + X_2}{2}$ and S^2 are

   independently distributed. Also show that $2S^2/\sigma^2$ follows χ^2 distribution with one d.f.

8. Let X_1, X_2 and X_3 be i.i.d. N (0, 1).

   Let, $\qquad Y_1 = \dfrac{(X_1 - X_2)}{\sqrt{2}}, \quad Y_2 = \dfrac{(X_1 + X_2 - 2X_3)}{\sqrt{6}}$ and $Y_3 = \dfrac{(X_1 + X_2 + X_3)}{\sqrt{3}}$

   Show that Y_1, Y_2, Y_3 are independently distributed. Hence or otherwise show that

   $\bar{X}$ and S^2 are independently distributed. Also show that $\bar{X} = \dfrac{Y_3}{\sqrt{3}} \to$ N (0, 1/3) and

   $3S^2 = Y_1^2 + Y_2^2$ follows chi-square distribution with 2 d.f.

9. Let $\bar{X}$ and S^2 be the mean an variance of a random sample of size 25 from N (3, 100)

   distribution. Compute P $(0 \le \bar{X} \le 6, 55.2 \le S^2 \le 145.6)$. **(P.U. May 2013)**

Exercise 4 (C)

Choose correct alternative out of (a) to (d) in following questions.

1. Let X_1, X_2, ..., X_{10} is a random sample from N (5, 9) then Var $(2\bar{X})$ is

   (a) $\dfrac{9}{10}$ (b) $\dfrac{10}{36}$ (c) $\dfrac{9}{5}$ (d) $\dfrac{36}{10}$

2. Suppose $Y_1, Y_2, ..., Y_{25}$ is a random sample from N (27, 50) with sample variance S^2. Then $\dfrac{S^2}{2}$ follows which of the distribution ?

   (a) Chi-square with 2 d.f. (b) Chi-square with 25 d.f.

   (c) Chi-square with 24 d.f. (d) Chi-square with 27 d.f.

3. Standard Error (S.E.) of a statistic T is given by

   (a) $E\sqrt{(T - E(T))^2}$ (b) $\sqrt{E[T - E(T)]^2}$

   (c) $E(T^2) - E(T)$ (d) $E\sqrt{T^2 - T}$

4. The random sample $X_1, X_2, ..., X_{16}$ drawn from N (μ, σ^2) gave mean 52. The probability distribution of $0.4 (52 - \mu)$ was observed to be t-distribution with 15 d.f. Hence the value of $\displaystyle\sum_{i=1}^{16} (X_i - \bar{X})^2$ is

   (a) 1500 (b) 1600 (c) 1400 (d) 1440

5. If $X_1, X_2, ..., X_n$ is a random sample from N (μ_1, σ^2) and $Y_1, Y_2, ..., Y_n$ is a random sample from N (μ_2, σ^2) then with usual notations $\dfrac{\bar{X} - \bar{Y} - (\mu_1 - \mu_2)}{\sqrt{s^2\left(\dfrac{1}{n_1} + \dfrac{1}{n_2}\right)}}$ where, s^2 = mean square of the pooled sample follows :

   (a) $n_1 + n_2$ d.f. (b) $n_1 + n_2 - 1$ d.f. (c) $(n_1 - 1)(n_2 - 1)$ d.f. (d) $n_1 + n_2 - 1$ d.f.

6. If $X_1, X_2 ... X_{n_1}$ and $Y_1, Y_2 ... Y_{n_2}$ denote random samples from N (μ_1, σ^2) and N (μ_2, σ^2) respectively, then $\dfrac{\sum (X_i - \bar{X})^2 / n_1 - 1}{\sum (Y_i - \bar{Y})^2 / n_2 - 1}$ has F-distribution with

   (a) $(n_2 - 1, n_1 - 1)$ d.f. (b) $(n_1 - 1, n_2 - 1)$ d.f.

   (c) (n_1, n_2) d.f. (d) (n_2, n_1) d.f.

7. If $X_1, X_2, ..., X_{16}$ is a random sample from G $\left(\dfrac{1}{2}, 1\right)$ distribution then standard error of sample mean $\bar{X}$ is

   (a) 2 (b) 4 (c) $\dfrac{1}{2}$ (d) $\dfrac{1}{4}$

8. If S^2 is a sample variance from a random sample from N (μ, σ^2) then the sampling distribution of $\dfrac{nS^2}{\sigma^2}$ is

   (a) $N\left(\mu, \dfrac{\sigma^2}{n}\right)$ (b) $N(\mu, n)$ (c) χ^2_{n-1} (d) χ^2_n

9. If $X_1, X_2, \ldots, X_n$ form a random sample from $N(\mu, \sigma^2)$ then the distribution of $\dfrac{\sum\limits_{i=1}^{n} X_i}{n}$ is

   (a) $N(\mu, \sigma^2)$ (b) $N(\mu, n\sigma^2)$ (c) $N\left(\mu, \dfrac{\sigma^2}{n^2}\right)$ (d) $N\left(\mu, \dfrac{\sigma^2}{n}\right)$

Exercise 4 (D)

State whether the following statements are true or false.

1. If $X_1, X_2, \ldots, X_{25}$ is a random sample from $N(35, 100)$ then $\text{Var}\left(\dfrac{\sum\limits_{i=1}^{25} X_i}{25}\right)$ is 2.

2. Let $X_1, X_2, \ldots, X_{16}$ be a random sample from $N(27, 100)$ with sample variance S^2 then $0.16\, S^2$ follows chi-square distribution with 15 d.f.

3. Statistic is a function of sample values involving unknown population parameter.

4. Statistic is a random variable while parameter is a constant.

5. Suppose $X_1, X_2, \ldots, X_n$ is a random sample from a population with unknown parameter θ and T is a statistic based on the sample values. Then $S{\cdot}E(T) = E(T) = \theta$.

6. If $X_1, X_2, \ldots, X_n$ is a random sample from $N(\mu, \sigma^2)$ then sample mean $\bar{X}$ and sample variance S^2 are not independent.

7. If $X_1, X_2, \ldots, X_{25}$ is a random sample from $N(50, \sigma^2)$ then the probability distribution of

$$\frac{5(\bar{X} - 50)}{\sqrt{\dfrac{\sum(X_i - \bar{X})^2}{24}}} \quad \text{is t with 24 d.f.}$$

8. If $X_1, X_2, \ldots X_{n_1}$ is a random sample from $N(\mu_1, \sigma^2)$ and $Y_1, Y_2 \ldots Y_{n_2}$ is a random sample from $N(\mu_2, \sigma^2)$ then with usual notations $\dfrac{\bar{X} - \bar{Y}}{\sqrt{s^2\left(\dfrac{1}{n_1} + \dfrac{1}{n_2}\right)}}$ where, $s^2 = $ mean square of pooled sample follows t-distribution with $n_1 + n_2 - 2$. d.f.

Answers of Exercise 4 (B)

(1) $F_{9,9}$ (2) 0.573   (3) 0.1707   (9) 0.1   (11) 14.4, 46.08, 0.05.

Answers of Exercise 4 (C)

(1) d   (2) c   (3) b   (4) a   (5) d
(6) b   (7) a   (8) c   (9) d

Answers of Exercise 4 (D)

(1) False   (2) True   (3) False   (4) True
(5) False   (6) False   (7) True   (8) False

Chapter **5**...

Exact Tests

Jerzy Neyman ForMemRS[1] (April 16, 1894 – August 5, 1981), born **Jerzy Spława-Neyman**, was a Polish mathematician and statistician who spent the first part of his professional career in various institutions in Warsaw, Poland, and the second part at theUniversity of California, Berkeley. Neyman first introduced the modern concept of a confidence interval into statistical hypothesis testing[2] and co-devised null hypothesis testing (in collaboration with Egon Pearson).

Jerzy Neyman

Contents ...

5.0   Introduction

5.1   Tests based on t-distribution

5.2   Tests based on χ^2 distribution

5.3   Tests based on F distribution

Key Words :

Statistic, level of significance, p-value, two sided confidence interval for parameter, testing of hypothesis, null hypothesis, alternative hypothesis, random sample from a probability distribution, contingency table, goodness of fit, chi-square test, one-sample t-test, two sample t-test, paired t-test, f-test.

Objectives :

(1) To make students familiar with the statistical tests of hypothesis, widely used in practice along with the assumptions.

(2) To understand derivation of test-statistic is using (sampling) distributions like chi-square, t-distribution and F-distribution under the respective null hypothesis.

(3) To construct confidence interval for the unknown population parameter with the help of tests of significance.

(4) To apply the test based on chi-square, t and F distribution in real life situations in the various fields such as medicine, genetics clinical trials, industry, socio-economic surveys, biostatistics, epidemiology, agriculture engineering, commerce etc. and draw valid conclusions using them.

5.0 Introduction

In the other book, we have seen large sample (or approximate) tests for testing a hypothesis about population mean, population proportion and correlation coefficient. However in number of situations such as biological experiments, clinical trials etc. where exact tests can be used for testing hypothesis about a population parameter.

In large sample tests, the distribution of test statistic Z is approximately taken as standard normal when the sample is of large size, which seems to be reasonable.

In small sample tests the statistic follows exact sampling distribution viz. χ^2, t or F for sample of any size. Hence small sample tests are regarded as exact tests. The procedure of testing a null hypothesis H_0 in small sample tests is almost identical to that of large sample test. For the sake of convenience, we divide these tests into three sets according to the distribution of test statistic as follows :

Tests based on

1. t-distribution
2. χ^2-distribution
3. F-distribution

5.1 Tests Based on t-Distribution

[A] Test for population mean.

Let $x_1, x_2, \ldots x_i, \ldots, x_n$ be a random sample of size n from a normal population with mean μ and variance σ^2. We desire to test $H_0 : \mu = \mu_0$ against $H_1 : \mu \neq \mu_0$ where μ_0 represents a specified value of population mean μ.

If the population variance σ^2 is known then the test statistic. Under $H_0 : \mu = \mu_0$ is,

$$\frac{\overline{X} - \mu_0}{\dfrac{\sigma}{\sqrt{n}}} \longrightarrow N(0, 1) \qquad \qquad \ldots (1)$$

The difficulty in testing H_0 arises when σ^2 is unknown. We shall discuss following two cases when σ^2 is *unknown*.

(i) $H_0 : \mu = \mu_0$ against $H_1 : \mu \neq \mu_0$.

(ii) $H_0 : \mu = \mu_0$ against $H_1 : \mu > \mu_0$ or $\mu < \mu_0$

Case (i) : $H_0 : \mu = \mu_0$ against $H_1 : \mu \neq \mu_0$

In order to overcome the difficulty that σ^2 is unknown, we propose the statistics U and V as follows :

$$U = \frac{\overline{x} - \mu}{\dfrac{\sigma}{\sqrt{n}}} \quad \text{and} \quad V = \frac{\Sigma (x_i - \overline{x})^2}{\sigma^2}$$

Then $U \sim> N(0, 1)$ and $V \sim> \chi^2$ distribution with $(n - 1)$ degrees of freedom. Moreover U and V are independent varieties. Hence, we can define statistic as $t = \dfrac{U}{\sqrt{\dfrac{V}{n-1}}}$ which has t-distribution with $(n - 1)$ degrees of freedom.

$$\therefore \quad t = \frac{U}{\sqrt{\dfrac{V}{n-1}}} = \frac{\dfrac{\bar{x} - \mu}{\dfrac{\sigma}{\sqrt{n}}}}{\sqrt{\dfrac{\sum (x_i - \bar{x})^2}{\sigma^2} \div (n-1)}}$$

$$= \frac{\bar{x} - \mu}{\dfrac{s}{\sqrt{n}}} \qquad \text{where, } s^2 = \frac{\sum (x_i - \bar{x})^2}{n-1} \qquad \ldots (2)$$

$$\therefore \quad \text{Under } H_0 : \mu = \mu_0, \ t = \frac{\bar{x} - \mu_0}{\dfrac{s}{\sqrt{n}}} \sim> \text{t-distribution with } (n-1) \text{ degrees of freedom.}$$

Thus we obtain test statistic which is free from unknown parameter σ.

Since, the alternative hypothesis $H_1 : \mu \neq \mu_0$ is two sided, the rejection region is $|t_{n-1}| \geq t_{n-1;\ \alpha/2}$ where, value $t_{n-1;\ \frac{\alpha}{2}}$ is such that $P\{|t_{n-1}| \geq t_{n-1;\ \alpha/2}\} = \alpha$ This value is also shown in the Fig. 5.1.

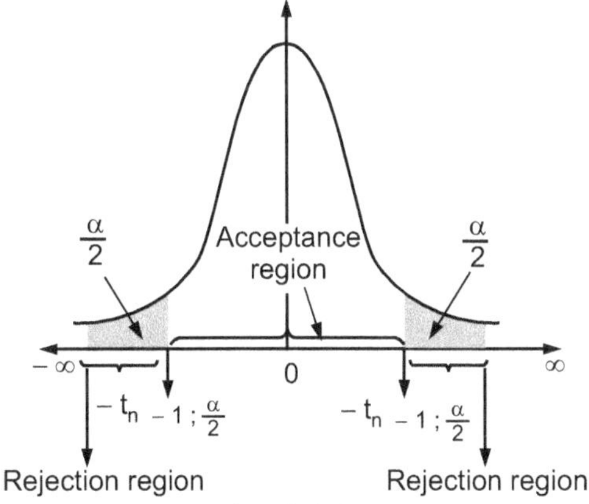

Fig. 5.1 (a)

So, the decision rule is, reject H_0 at level of significance α, if $|t_{n-1}| \geq t_{n-1;\ \alpha/2}$ and accept H_0 otherwise. Then the conclusion about the population mean can be drawn accordingly.

Decision using p-value : We note that p-value = P {critical region}. If p value is level of significance, we accept H_0 and reject H_0 otherwise.

Remarks : (I) If σ is known, we use test statistic $\dfrac{\overline{X} - \mu_0}{\sigma/\sqrt{n}}$. It follows N(0, 1) distribution,

where as if σ is unknown, we use test statistic $\dfrac{\overline{X} - \mu_0}{s/\sqrt{n}}$. Apparently we may think that the

unknown σ is replaced by s but it is not so which is clear from the above discussion.

(II) For using any test based on t-distribution, one has to ensure that the observations in the parent population (population from which sample is drawn) follow normal distribution.

(III) We can get rough idea about the acceptance of $H_0 : \mu = \mu_0$ using box plot also. Before conducting actual test, we draw box plot.

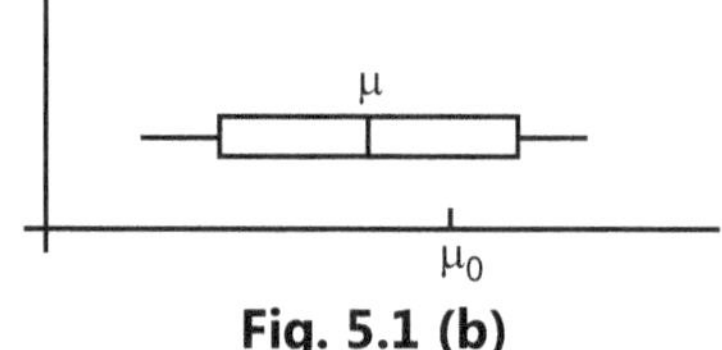

Fig. 5.1 (b)

Confidence Interval for Parameter μ : Let $x_1, x_2, ..., x_n$ be a random sample from the probability density function $f(x, \mu)$ where, μ is the parameter of the distribution. The interval (T_1, T_2) is called $100 (1 - \alpha)$ % confidence interval (C.I.) for μ if $P[T_1 < \mu < T_2] = 1 - \alpha$ where, T_1 and T_2 are some functions sample observations. In this case $1 - \alpha$ is called confidence coefficient. In particular if $P[T_1 < \mu < T_2] = 0.99$ then (T_1, T_2) is called 99% confidence interval for parameter μ or confidence interval with confidence coefficient 0.99. Similarly, for 95% confidence interval for parameter μ, we must have $P[T_1 < \mu < T_2] = 0.95$.

T_1 is called lower confidence bound or lower confidence limit while T_2 is called upper confidence bound or upper confidence limit. The difference between upper confidence limit and lower confidence limit $(T_2 - T_1)$ is called length of confidence interval. The confidence intervals can be obtained using the corresponding critical region in testing of hypotheses.

Confidence interval for population mean (μ) when population S.D. (σ) is known :

We know that if one wants to test $H_0 : \mu = \mu_0$ against $H_1 : \mu \neq \mu_0$ when σ is known

$$Z = \frac{\overline{X} - \mu}{\sigma/\sqrt{n}} \rightarrow N(0, 1)$$

$\therefore \qquad P[|Z| \geq z_{\alpha/2}] = \alpha$

OR $P[|Z| \leq z_{\alpha/2}] = 1 - \alpha$

$$\therefore \qquad P\left[\left|\frac{\overline{X} - \mu}{\sigma/\sqrt{n}}\right| \leq z_{\alpha/2}\right] = 1 - \alpha$$

$$P\left[|\overline{X} - \mu| \leq z_{\alpha/2}\frac{\sigma}{\sqrt{n}}\right] = 1 - \alpha$$

Let $t = z_{\alpha/2}\dfrac{\sigma}{\sqrt{n}}$ then

$$P\,[|\bar{X} - \mu| \le t] = 1 - \alpha$$

$\therefore \quad P\,[\bar{X} - t < \mu < \bar{X} + t] = 1 - \alpha \quad \text{where, } t = z_{\alpha/2}\dfrac{\sigma}{\sqrt{n}}$

Thus, when population S.D. σ is known, $100\,(1 - \alpha)\,\%$ confidence interval (C.I.) for population mean is given by

$(\bar{X} - t,\ \bar{X} + t)$ That is $\left(\bar{X} - z_{\alpha/2}\dfrac{\sigma}{\sqrt{n}},\ \bar{X} + z_{\alpha/2}\dfrac{\sigma}{\sqrt{n}}\right)$

Here lower limit of C.I. $= T_1 = \bar{X} - t = \bar{X} - z_{\alpha/2}\dfrac{\sigma}{\sqrt{n}}$.

Upper limit of C.I. $= T_2 = \bar{X} + t = \bar{X} + z_{\alpha/2}\dfrac{\sigma}{\sqrt{n}}$.

Thus, confidence limits are $\bar{X} \mp t$ or $\bar{X} \mp z_{\alpha/2}\dfrac{\sigma}{\sqrt{n}}$.

In this case length of confidence interval

$$= \text{Upper confidence bound} - \text{Lower confidence bound}$$

$$= 2t = 2\, z_{\alpha/2}\dfrac{\sigma}{\sqrt{n}} = \text{constant}$$

In particular when $\alpha = 0.05$, $t = z_{\alpha/2}\dfrac{\sigma}{\sqrt{n}} = Z_{0.025}\dfrac{\sigma}{\sqrt{n}}$.

$\therefore$ 95% confidence interval for population mean μ when σ is known;

$$\left(\bar{X} - Z_{0.025}\dfrac{\sigma}{\sqrt{n}},\ \bar{X} + Z_{0.025}\dfrac{\sigma}{\sqrt{n}}\right) \text{ i.e. } \left(\bar{X} - 1.96\dfrac{\sigma}{\sqrt{n}},\ \bar{X} + 1.96\dfrac{\sigma}{\sqrt{n}}\right).$$

Similarly, for $\alpha = 0.01$, $t = z_{\alpha/2}\dfrac{\sigma}{\sqrt{n}} = Z_{0.005}\dfrac{\sigma}{\sqrt{n}} = 2.58\dfrac{\sigma}{\sqrt{n}}$.

$\therefore$ 99% confidence interval for population mean μ when σ is known will be

$$\left(\bar{X} - Z_{0.005}\dfrac{\sigma}{\sqrt{n}},\ \bar{X} + Z_{0.005}\dfrac{\sigma}{\sqrt{n}}\right) \text{ i.e. } \left(\bar{X} - 2.58\dfrac{\sigma}{\sqrt{n}},\ \bar{X} + 2.58\dfrac{\sigma}{\sqrt{n}}\right)$$

This means that in 99 cases out of 100, our population mean will lie within above interval.

Confidence Interval for population mean (μ) when population S.D. (σ) is unknown.

We note that while testing $H_0 : \mu = \mu_0$ against $H_1 : \mu \ne \mu_0$ when σ is unknown, we get

$$P\left\{|t_{n-1}| \ge t_{n-1;\,\alpha/2}\right\} = 1 - \alpha. \text{ Let } t' = t_{n-1;\,\alpha/2}$$

$\therefore \quad P\,\{|t_{n-1}| \ge t'\} = 1 - \alpha$

i.e. $\quad P\left\{\left|\dfrac{\bar{X} - \mu}{s/\sqrt{n}}\right| \ge t'\right\} = 1 - \alpha$

$$P\left\{|\bar{X} - \mu| \geq t'\frac{s}{\sqrt{n}}\right\} = 1 - \alpha$$

$$\therefore \quad P\left\{\bar{X} - t'\frac{s}{\sqrt{n}} \leq \mu \leq \bar{X} + t'\frac{s}{\sqrt{n}}\right\} = 1 - \alpha$$

Hence, in this case $100(1 - \alpha)\%$ C.I. for population mean μ when population S.D. is unknown will be

$$\left(\bar{X} - t'\frac{s}{\sqrt{n}}, \bar{X} + t'\frac{s}{\sqrt{n}}\right) \quad \text{where } t' = t_{n-1;\, \alpha/2}.$$

$$\therefore \quad \text{Lower limit of C.I.} = T_1 = \bar{X} - t'\frac{s}{\sqrt{n}}$$

$$\text{Upper limit of C.I.} = T_2 = \bar{X} + t'\frac{s}{\sqrt{n}}$$

$$\text{Length of C.I.} = 2t'\frac{s}{\sqrt{n}} = 2\, t_{n-1;\, \alpha/2}\frac{s}{\sqrt{n}} \quad \text{(It is a variable)}$$

In particular if $\alpha = 0.05$, $t' = t_{n-1;\, 0.025}$.

Then 95% C.I. for population mean μ in this case is

$$\left(\bar{X} - t'\frac{s}{\sqrt{n}}, \bar{X} + t'\frac{s}{\sqrt{n}}\right) = \left(\bar{X} - t_{n-1;\, 0.025}\frac{s}{\sqrt{n}}, \bar{X} + t_{n-1;\, 0.025}\frac{s}{\sqrt{n}}\right)$$

Similarly, when $\alpha = 0.01$, we have $t' = t_{n-1;\, 0.005}$.

$\therefore$　Confidence interval (C.I.) for μ (when σ is unknown) with confidence coefficient 0.99 is

$$\left(\bar{X} - t_{n-1;\, 0.005}\frac{s}{\sqrt{n}}, \bar{X} + t_{n-1;\, 0.005}\frac{s}{\sqrt{n}}\right)$$

Thus, confidence limits depend on s, n and table values of t-distribution.

There are following two uses of constructing confidence interval (C.I.) in

(i) Testing of hypothesis : We can accept the null hypothesis $H_0 : \mu = \mu_0$ if the value of μ_0 lies in the confidence interval and Reject it otherwise at given level of significance α. In this case critical region is viewed as sample space.

(ii) Estimation of parameter : We get interval within which value of parameter lies with specified confidence coefficient based on sample mean. In this case critical region is viewed as parameter space.

Case (ii) : $H_0 : \mu = \mu_0$ against $H_1 : \mu < \mu_0$ or $\mu > \mu_0$. In this case we use the same test statistic as in case (i). If $H_1 : \mu < \mu_0$, then the critical region at level of significance α is $t_{n-1} \leq -t_{n-1;\, \alpha}$. It is shaded region as shown in Fig. 5.2.

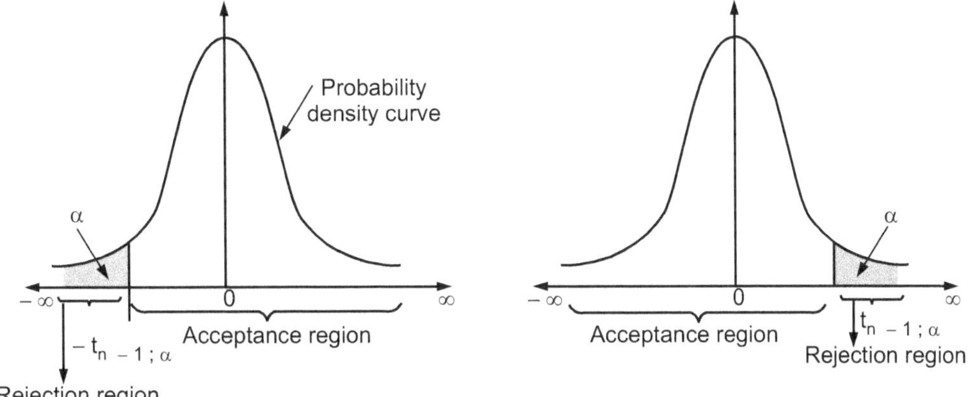

Fig. 5.2   **Fig. 5.3**

On the other hand if $H_1 : \mu > \mu_0$ then critical region at level of significance α is $t_{n-1} \geq t_{n-1;\alpha}$. It is shaded region as shown in Fig. 5.3.

[B] Testing equality of means of two populations :

Suppose $X_1, X_2, \ldots X_i, \ldots, X_{n_1}$ be a random sample of size n_1 drawn from normal population with parameters μ_1 and σ_1^2. Also $Y_1, Y_2 \ldots Y_i, \ldots, Y_{n_2}$ be a random sample of size n_2 drawn from another population with parameter μ_2 and σ_2^2. If we assume that the two samples are independent and population variances σ_1^2 and σ_2^2 are unknown but *equal* i.e. $\sigma_1^2 = \sigma_2^2 = \sigma^2$, then a test based on t-distribution can be used to test $H_0 : \mu_1 = \mu_2$ described below.

Notation :

$\overline{x}$: Arithmetic mean of sample of size n_1 drawn from 1st population.

$\overline{y}$: Arithmetic mean of sample of size n_2 drawn from 2nd population.

$$s^2 = \text{Mean square for pooled samples}$$

$$= \frac{\displaystyle\sum_{i=1}^{n_1} (x_i - \overline{x})^2 + \sum_{i=1}^{n_2} (y_i - \overline{y})^2}{n_1 + n_2 - 2}$$

$$= \frac{\left[\displaystyle\sum_{i=1}^{n_1} x_i - n_1 \overline{x}^2\right] + \left[\displaystyle\sum_{i=1}^{n_2} y_i^2 - n_2 \overline{y}^2\right]}{n_1 + n_2 - 2}$$

We want to obtain a test statistic based on observations in the samples drawn from the two populations.

Let,

$$U_1 = \frac{\overline{x} - \mu_1}{\dfrac{\sigma}{\sqrt{n_1}}} \; , \quad V_1 = \frac{\displaystyle\sum_{i=1}^{n_1} (x_i - \overline{x})^2}{\sigma^2}$$

and

$$U_2 = \frac{\overline{y} - \mu_2}{\dfrac{\sigma}{\sqrt{n_2}}} \; , \quad V_2 = \frac{\displaystyle\sum_{i=1}^{n_2} (y_i - \overline{y})^2}{\sigma^2}$$

Then $U_1 \to N(0, 1)$, $U_2 \to N(0, 1)$, $V_1 \to \chi^2_{n_1 - 1}$ and $V_2 \to \chi^2_{n_2 - 1}$

U_1, U_2, V_1 and V_2 are all independently distributed random variables.

$$\therefore \quad U_1 - U_2 = \frac{(\overline{x} - \overline{y}) - (\mu_1 - \mu_2)}{\sigma \sqrt{\dfrac{1}{n_1} + \dfrac{1}{n_2}}} \to N(0, 1) \qquad \ldots \text{(i)}$$

and by additive property of chi-square random variables, we have,

$$\text{Suppose, } V = V_1 + V_2 = \frac{\displaystyle\sum_{i=1}^{n_1} (x_i - \overline{x})^2 + \sum_{i=1}^{n_2} (y_i - \overline{y})^2}{\sigma^2} \to \chi^2_{n_1 + n_2 - 2}$$

$$\therefore \quad V = \frac{\displaystyle\sum_{i=1}^{n_1} (x_i - \overline{x})^2 + \sum_{i=1}^{n_2} (y_i - \overline{y})^2}{\sigma^2} \to \chi^2_{n_1 + n_2 - 2} \qquad \ldots \text{(ii)}$$

Now, consider the statistic

$$t = \frac{U_1 - U_2}{\sqrt{\dfrac{V}{n_1 + n_2 - 2}}}$$

From equation (i) and (ii), we have $t \to$ t-distribution with $(n_1 + n_2 - 2)$ degrees of freedom,

Hence,

$$t = \frac{(\overline{x} - \overline{y}) - (\mu_1 - \mu_2)}{\sigma \sqrt{\dfrac{1}{n_1} + \dfrac{1}{n_2}}} \times \frac{1}{\sqrt{\dfrac{\displaystyle\sum_{i=1}^{n_1} (x_i - \overline{x})^2 + \sum_{i=1}^{n_2} (y_i - \overline{y})^2}{\sigma^2 (n_1 + n_2 - 2)}}}$$

$$= \frac{\left(\overline{x} - \overline{y}\right) - (\mu_1 - \mu_2)}{\left[\sqrt{\dfrac{\sum\limits_{i=1}^{n_1}(x_i - \overline{x})^2 + \sum\limits_{i=1}^{n_2}(y_i - \overline{y})^2}{(n_1 + n_2 - 2)}}\right]\sqrt{\dfrac{1}{n_1} + \dfrac{1}{n_2}}}$$

$$= \frac{\overline{x} - \overline{y} - (\mu_1 - \mu_2)}{s\sqrt{\dfrac{1}{n_1} + \dfrac{1}{n_2}}} \qquad \text{where, } s \text{ is positive square root of } s^2$$

$\therefore$ If $H_0 : \mu_1 = \mu_2$ is true,

$$t = \frac{\overline{x} - \overline{y}}{s\sqrt{\dfrac{1}{n_1} + \dfrac{1}{n_2}}} \quad \text{has } t\text{-distribution with } (n_1 + n_2 - 2) \text{ degrees of freedom.}$$

Hence, the critical region for testing $H_0 : \mu_1 = \mu_2$ against $H_1 : \mu_1 \neq \mu_2$ at level of significance α contains of all the values of statistic for which $|t| \geq t_{n_1 + n_2 - 2;\ \alpha/2}$.

Thus, we reject H_0 at level of significance α if $|t| \geq t_{n_1 + n_2 - 2;\ \alpha/2}$ and H_0 otherwise.

It follows that for testing $H_0 : \mu_1 = \mu_2$ against $H_1 : \mu_1 > \mu_2$ the critical region level of significance is $t \geq t_{n_1 + n_2 - 2;\ \alpha}$ on the other hand in order to test $H_0 : \mu_1 = \mu_2$ against $H_1 : \mu_1 < \mu_2$, it is given by $t \leq - t_{n_1 + n_2 - 2;\ \alpha}$.

Remarks :

1. Suppose s_1^2 and s_2^2 are mean squares for the samples drawn from first and second population respectively.

$$\therefore \quad s_1^2 = \frac{\sum\limits_{i=1}^{n_1}\left(x_i - \overline{x}\right)^2}{n_1 - 1} = \frac{\sum\limits_{i=1}^{n_1} x_i^2 - n_1 \overline{x}^2}{n_1 - 1}$$

$$\text{and} \quad s_2^2 = \frac{\sum\limits_{i=1}^{n_2}\left(y_i - \overline{y}\right)^2}{n_2 - 1} = \frac{\sum\limits_{i=1}^{n_2} y_i^2 - n_2 \overline{y}^2}{n_2 - 1}$$

Then s^2 can be expressed in terms of s_1^2 and s_2^2 as $s^2 = \dfrac{(n_1 - 1) s_1^2 + (n_2 - 1) s_1^2}{n_1 + n_2 - 2}$

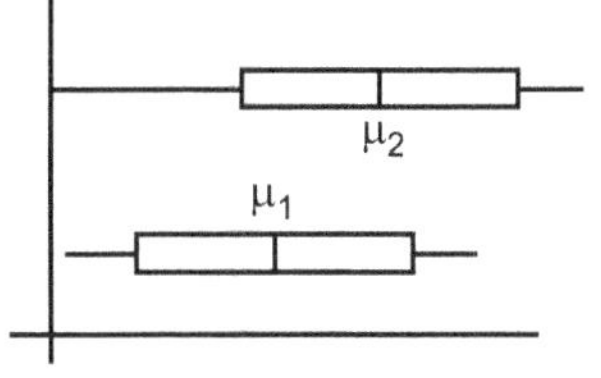

Fig. 5.3 (a)

2. Let S_1^2 and S_2^2 represent the variances of samples of sizes n_1 and n_2 respectively.

$$\therefore \qquad S_1^2 = \frac{\sum\limits_{i=1}^{n_1} \left(x_i - \bar{x}\right)^2}{n_1}$$

$$\text{and} \qquad S_2^2 = \frac{\sum\limits_{i=1}^{n_2} \left(y_i - \bar{y}\right)^2}{n_2}$$

$$\text{then} \qquad s^2 = \frac{n_1 S_1^2 + n_2 S_2^2}{n_1 + n_2 - 2}$$

Remark (3) : With the help of box plot we can ensure $\mu_1 = \mu_2$

Remark (4) : Confidence interval for difference between population means :

While testing $H_0 : \mu_1 = \mu_2$ against $H_1 : \mu_1 \neq \mu_2$ we get

$$P\left[|t| \geq t_{n_1 + n_2 - 2;\, \frac{\alpha}{2}}\right] = \alpha. \text{ Let } t^* = t_{n_1 + n_2 - 2;\, \alpha/2}.$$

$$\therefore \qquad P\left[|t| \geq t^*\right] = \alpha$$

OR $\qquad\qquad P\left[|t| \leq t^*\right] = 1 - \alpha$

$$\therefore \qquad P\left[\left|\frac{\bar{X} - \bar{Y} - (\mu_1 - \mu_2)}{s\sqrt{\dfrac{1}{n_1} + \dfrac{1}{n_2}}}\right| \leq t^*\right] = 1 - \alpha$$

$$P\left[|\bar{X} - \bar{Y} - (\mu_1 - \mu_2)| \leq t^* s\sqrt{\frac{1}{n_1} + \frac{1}{n_2}}\right] = 1 - \alpha$$

$$\therefore \quad P\left[(\bar{X} - \bar{Y}) - t^* s\sqrt{\frac{1}{n_1} + \frac{1}{n_2}} \leq \mu_1 - \mu_2 \leq (\bar{X} - \bar{Y}) + t^* s\sqrt{\frac{1}{n_1} + \frac{1}{n_2}}\right] = 1 - \alpha$$

Hence, $100 (1 - \alpha) \%$ C.I. for difference between two population means $\mu_1 - \mu_2$ is given by

$$\left[(\bar{X} - \bar{Y}) - t^* s\sqrt{\frac{1}{n_1} + \frac{1}{n_2}},\ (\bar{X} - \bar{Y}) + t^* s\sqrt{\frac{1}{n_1} + \frac{1}{n_2}}\right]$$

where $t^* = t_{n_1 + n_2 - 2;\, \alpha/2}$.

$$\therefore \quad \text{Lower confidence limit} = (\bar{X} - \bar{Y}) - t^* s\sqrt{\frac{1}{n_1} + \frac{1}{n_2}}$$

$$\text{Upper confidence limit} = (\bar{X} - \bar{Y}) + t^* s\sqrt{\frac{1}{n_1} + \frac{1}{n_2}}$$

Thus, confidence limits are $(\bar{X} - \bar{Y}) \pm t^* s\sqrt{\dfrac{1}{n_1} + \dfrac{1}{n_2}}$.

Length of C.I. = Upper confidence limit – Lower confidence limit

$$= 2t^* s \sqrt{\frac{1}{n_1} + \frac{1}{n_2}}$$

$\therefore$ In particular if $\alpha = 0.05$,

$$t^* = t_{n_1 + n_2 - 2;\ 0.005}$$

$\therefore$ 95% confidence interval for difference between population means is

$$\left[(\bar{X} - \bar{Y}) - t_{n_1 + n_2 - 2;\ 0.005}\ s \sqrt{\frac{1}{n_1} + \frac{1}{n_2}}\ ,\ (\bar{X} - \bar{Y}) + t_{n_1 + n_2 - 2;\ 0.005}\ s \sqrt{\frac{1}{n_1} + \frac{1}{n_2}} \right]$$

Similarly, when $\alpha = 0.01$, $t^* = t_{n_1 + n_2 - 2;\ 0.005}$.

$\therefore$ 99% C.I. for difference between two population means is

$$\left[(\bar{X} - \bar{Y}) - t_{n_1 + n_2 - 2;\ 0.005}\ s \sqrt{\frac{1}{n_1} + \frac{1}{n_2}}\ ,\ (\bar{X} - \bar{Y}) + t_{n_1 + n_2 - 2;\ 0.005}\ s \sqrt{\frac{1}{n_1} + \frac{1}{n_2}} \right]$$

Summary : Confidence Intervals :

Parameter of normal population	Limits for 100 $(1 - \alpha)$ % C.I.		Remark
	Lower limit	**Upper limit**	
Population mean μ when σ is known	$\bar{X} - t$	$\bar{X} + t$	$t = z_{\alpha/2}\ \dfrac{\sigma}{\sqrt{n}}$
Population mean μ when σ is unknown	$\bar{X} - t'\ \dfrac{s}{\sqrt{n}}$	$\bar{X} + t'\ \dfrac{s}{\sqrt{n}}$	$t' = t_{n-1;\ \alpha/2}$
Difference between two population means $\mu_1 - \mu_2$	$\bar{X} - t^* s \sqrt{\dfrac{1}{n_1} + \dfrac{1}{n_2}}$	$\bar{X} + t^* s \sqrt{\dfrac{1}{n_1} + \dfrac{1}{n_2}}$	$t^* = t_{n_1 + n_2 - 2;\ \alpha/2}$

Limitations :

1.   This test is carried out under the assumption that the samples are drawn from independent normal population.

2.   If the populations from which the samples are drawn do not have equal variances $\left(\sigma_1^2 \neq \sigma_2^2 \right)$ then the above test cannot be used.

[C]   Paired t-Test

Sometimes following situations where the two samples are dependent occur in practice :

1.   We are interested in testing whether a training is effective in improving average performance of personnel.

2.   A producer may desire to know how far maintenance or overhauling of machine results into better functioning.

3.   A bulky person may be thinking to join a health club for reducing his weight. He desires to test that after joining the health club the average weight of participants gets reduced significantly or not. For this purpose he can collect the data on the weights of participants before and after joining the club and then take decision whether to join the club or not. In such cases paired t-test is used. It is carried out as follows :

Let $\{(X_i, Y_i)\ i = 1, 2, \ldots, n\}$ be a random sample from bivariate normal population with mean of X as μ_x and that of Y is μ_y, and unknown variables σ_x^2 and σ_y^2 .

Let $d_i = y_i - x_i$ for $i = 1, 2, \ldots, n$

Note that $d_i \to N\left(\mu_d, \sigma_d^2\right)$ where, $\mu_d = \mu_Y - \mu_X$.

In paired t-test, we want to test $H_0 : \mu_d = 0$ against $H_1 : \mu_d \neq 0$ or $\mu_d < 0$ or $\mu_d > 0$.

Thus, the test reduces to the test for single sample discussed in **5.1 (A)**.

Suppose,
$$\overline{d} = \frac{\Sigma d_i}{n}$$

and
$$s^2 = \frac{\Sigma(d_i - \overline{d})^2}{n - 1} = \frac{\Sigma d_i^2 - n\,\overline{d}^2}{n - 1}$$

Then under $H_0 : \mu_d = 0$, the test statistic

$$t = \frac{\overline{d}}{\dfrac{s}{\sqrt{n}}} = \frac{\overline{d}\sqrt{n}}{s}$$

has t-distribution with $(n - 1)$ degrees of freedom.

Hence, the critical regions for testing $H_0 : \mu_d = 0$ at level of significance α for different types of alternative hypotheses are as follows :

Type of H_1	Critical region Calculated value Table value
$H_1 : \mu_d \neq 0$	$\lvert t_{n-1} \rvert \geq t_{n-1;\,\frac{\alpha}{2}}$
$H_1 : \mu_d > 0$	$t_{n-1} \geq t_{n-1;\,\alpha}$
$H_1 : \mu_d < 0$	$t_{n-1} \leq -t_{n-1;\,\alpha}$

Note :

Here observations in two samples are dependent unlike usual t-test in which the observations in the two samples are independent. The confidence interval in this can be obtained similarly.

Solved Examples

Example 5.1 : In order to start new S.T. bus to a certain remote village it is required to get the average fare of ₹ 400 daily. Reports on number of passengers for 21 days revealed that the average daily collection of fare from the passengers was ₹ 390 with standard deviation of ₹ 40. Do these data support the demand of people for starting new bus to the village ? [Use 5% level of significance]

Solution : Here, we want to test $H_0 : \mu = \mu_0 = 400$ against $H_1 : \mu < 400$.

∴ The test statistic is given by

$$t_{n-1} = \frac{\overline{X} - \mu_0}{\frac{s}{\sqrt{n}}}$$

where, $\overline{x}$ = sample mean $= 390$

and
$$s^2 = \frac{\Sigma (x_i - \overline{x})^2}{n-1} = \frac{n \text{ (sample variance)}}{n-1}$$

$$= \frac{(21)\,(1600)}{20} = 1680 \qquad\qquad (\because \text{S.D.} = 40)$$

∴ $s = 40.987803$

∴
$$t_{20} = \frac{390 - 400}{\frac{40.987803}{\sqrt{21}}} = \frac{(-10)\sqrt{21}}{40.987803}$$

$$= -1.1180342 \text{ [calculated value]}$$

$$t_{20;\,0.05} = 1.725 \text{ [table value]}$$

[Since H_1 is one sided see the value in statistical table corresponding to n = 20 and probability 0.1].

As $H_1 : \mu < \mu_0$ the decision rule is reject H_0 if t_{20} (calculated value) $< - t_{20;\,0.05} = -1.725$ and accept H_0 otherwise. Here $t_{20} = -1.1180342 > -1.725$. Hence we accept H_0 at 5 % level of significance.

Conclusion : These data support the demand of people for starting new bus to the village.

Second method using p-value : p-value = $P\,[t_{20} < -1.118] = P\,[t_{20} > 1.118] = 0.1384$

∴ p-value > level of significancee = 0.05 ∴ our decision is accept H_0 at 5% level of significance.

Example 5.2 : Suppose that sweets are sold in packages of fixed weight of the contents. The producer of the packages is interested in testing that average weight of contents in packages in 1 kg. Hence a random sample of 12 packages is drawn and their contents found (in kg) as follows : 1.05, 1.01, 1.04, 0.98, 0.96, 1.01, 0.97, 0.99, 0.98, 0.95, 0.97, 0.95.

(i)   Test normality using normal probability paper.

(ii) Using the above data what should he conclude about the average weight of contents in the packets ? [Use 5 % level of significance]

(iii) Construct 95% confidence interval for weight of contents in packages. Also find length of confidence interval.

Solution : (i) To test normality we arrange the observations in increasing order and plot them on normal probability paper.

$X_{(i)}$	i	$P = \dfrac{i - 0.5}{n} = \dfrac{i - 0.5}{12}$	100 P
0.95	2	0.125	12.50
0.95	2	0.125	12.50
0.96	3	0.2083	20.83
0.97	5	0.375	37.50
0.97	5	0.375	37.50
0.98	7	0.5417	54.17
0.98	7	0.5417	54.17
0.99	8	0.625	62.50
1.01	10	0.7917	79.17
1.01	10	0.7917	79.17
1.04	11	0.875	87.50
1.05	12	0.9583	95.83

Note : i denotes the number of obervations less than or equal to $X_{(i)}$.

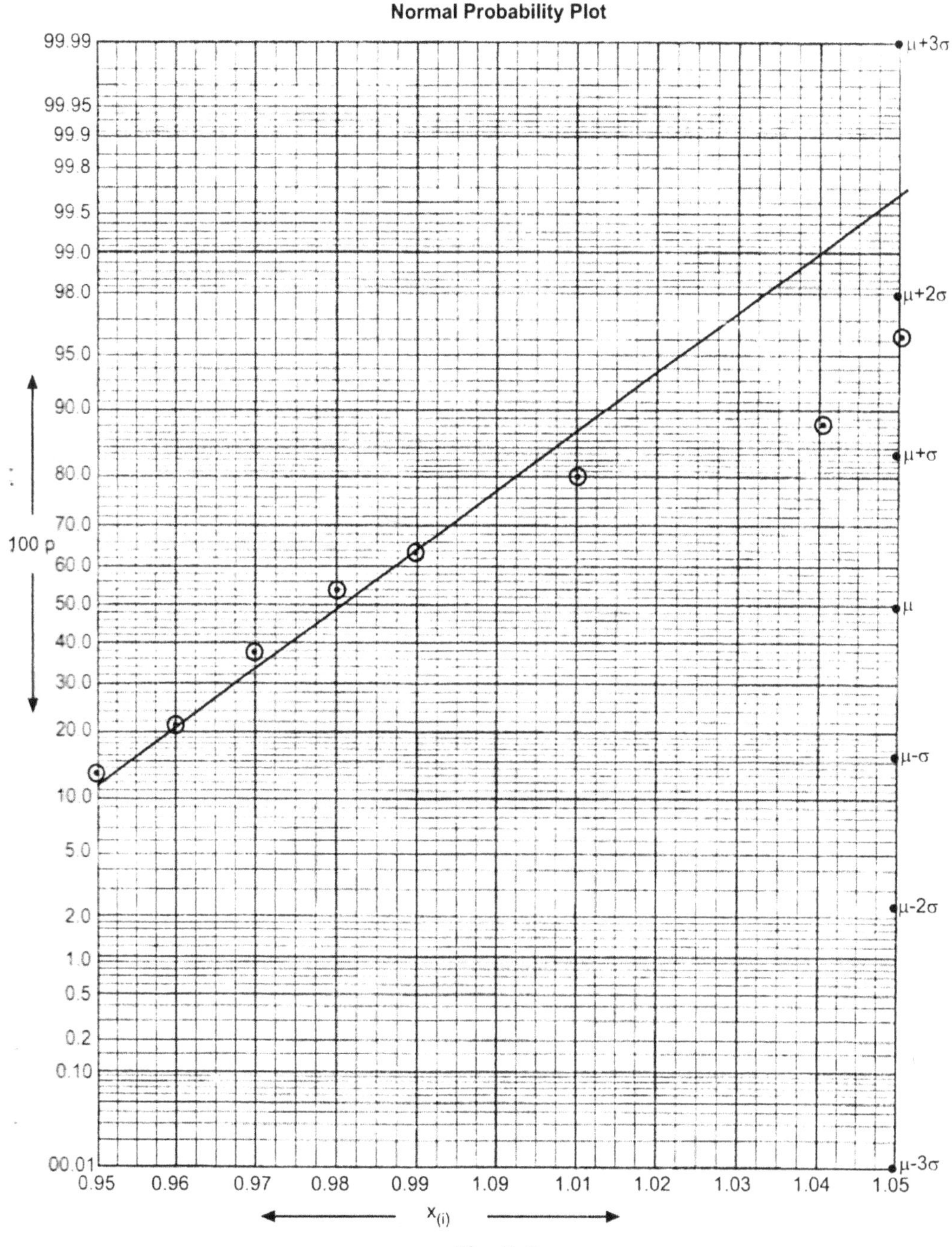

Fig. 5.4

Conclusion : From the graph the observations seem to come from normal distribution.

Let μ denote the average weight of contents in packages. Producer wants to test $H_0 : \mu = \mu_0 = 1$ against $H_1 : \mu \neq 1$.

Test statistic is given by

$$t_{n-1} = \frac{\bar{x} - \mu_0}{\dfrac{s}{\sqrt{n}}}$$

where, $\bar{x}$ = sample mean = $\dfrac{\Sigma x_i}{n}$

and $s^2 = \dfrac{\Sigma x_i^2 - n\bar{x}^2}{n-1}$

Here, $\bar{x}$ = 0.9883,

$$s^2 = \frac{11.7376 - 11.7208}{11} = 0.0015273$$

$\therefore$ s = 0.03908

$\therefore$ $t_{11} = \dfrac{(0.9883 - 1)\sqrt{12}}{0.03908} = -\dfrac{0.04053}{0.03908} = -1.0371$

$\therefore$ $|t_{11}|$ = 1.188145 [calculated value] < $t_{11; 0.025}$ = 2.201 [Table value]

$\therefore$ We accept H_0 at 5 % level of significance.

p-value = P [$|t_{11}|$ > 1.188] (two failed) = 0.1299 > level of significance = 0.05.

$\therefore$ Accept H_0

Confidence interval for population mean.

Here population S.D. is unknown

$\therefore$ 95% for population mean is $\left(\bar{X} - t' \dfrac{s}{\sqrt{n}}, \bar{X} + t' \dfrac{s}{\sqrt{n}}\right)$ where t' = $t_{n-1; \alpha/2}$.

Here α = 0.05, n = 12 $\therefore$ t' = $t_{11; 0.025}$ = 2.201 and $\bar{X}$ = 0.9883, s = 0.034112.

$\therefore$ 95% of confidence interval for mean weight of contents in packages is

$$\left[0.9883 - \frac{(2.0201)\,(0.034112)}{\sqrt{12}}, \; 0.9883 + \frac{(2.201)\,(0.034112)}{\sqrt{12}}\right]$$

= 0.9883 – 0.26009, 0.9883 + 0.26009]

= [0.72821, 1.24839]

Length of confidence level = 2 (0.26009) = 0.52018

Remark : The value of μ_0 = 1. It lies in the interval (0.72821, 1.24839). It supports the decision of accepting H_0.

Conclusion : The producer should conclude that the average weight of contents of package may be taken as 1 kg.

Solution using R software : We check whether the sample is from N (μ, σ^2) which is a assumption for using t test.

To test the normality of data we use Shapiro test as follows given by R software command, $\boxed{\text{Shapiro.test(x)}}$: where x is a vector containing observations.

R commands are

```
>x=c(1.05,1.01,1.04,0.98,0.96,1.01,0.97,0.99,0.98,0.95,0.97,0.95)
>Shapiro.test (x)
```

Which gives output as

Shapiro-Wilk normally test

data : x

W = 0.9144, p-value = 0.2428

From the output since the p-value is greater than level os significance $\alpha = 0.05$ we accept hypothesis of normality of data. Hence observations seem to come from normal distribution.

For testing $H_0 : \mu = \mu_0 = 1$ against $H_1 : \mu \neq \mu_0$ we use commands.

```
>x=c(1.05,1.01,1.04,0.98,0.96,1.01,0.97,0.99,0.98,0.95,0.97,0.95)
>t.test(x,mu=1)
```

Which gives output

One sample t-test

data : x

t = $-$ 1.2253, df = 11, p-value = 0.2460

alternative hypothesis : true mean is not equal to 1

95 percent confidence interval :

0.967377   1.009290

sample estimates :

mean of x

0.9883333

From the output since the p-value is greater than level of significance $\alpha = 0.05$ we accept hypothesis.

Example 5.3 : The ability of brakes of two new types of cars was tested by driving the cars at the speed of 60 miles per hour and then applying the breaks. The distances (in feet) required to stop the cars were noted. The results were as follows :

Types of the Car	I	II
Number of cars tested	12	15
Average distance (in inches) required to stop	7.8	9.6
S.D. of the distance	2.2	3.5

(i) From above results, will you conclude that average distances required to stop using breaks of two new types of cars are equal ?

(ii) Find 99% confidence interval for difference between average distances required to stop the cars of the two types.

Solution :

Let, μ_1 : Average distance required to stop using breaks of Cars of type I.

μ_2 : Average distance required to stop using breaks of Cars of type II.

Here, we want to test $H_0 : \mu_1 = \mu_2$ against $H_1 : \mu_1 \neq \mu_2$. The test statistic is given by,

$$t_{n_1 + n_2 - 2} = \frac{\overline{x} - \overline{y}}{s\sqrt{\dfrac{1}{n_1} + \dfrac{1}{n_2}}}$$

where, $$s^2 = \frac{[\Sigma(x_i - \overline{x})^2 + \Sigma(y_i - \overline{y})^2]}{n_1 + n_2 - 2}$$

$$= \frac{n_1 S_1^2 + n_2 S_2^2}{n_1 + n_2 - 2}$$

where, S_1^2 = Variance of observations in the first sample

$$= (2.2)^2$$

S_2^2 = Variance of observations in the second sample

$$= (3.5)^2$$

$\therefore$ $$s^2 = \frac{(12)(2.2)^2 + (15)(3.5)^2}{12 + 15 - 2}$$

$$= \frac{241.83}{25} = 9.6732$$

$\therefore$ $$s = 3.1101768$$

$\therefore$ $$t_{25} = \frac{7.8 - 9.6}{3.1101768\sqrt{\dfrac{1}{12} + \dfrac{1}{15}}}$$

$$= \frac{-1.8}{1.2045658} = -1.4943143$$

$\therefore$ $$|t_{25}| = 1.4943143 \qquad \text{[Calculated value]}$$

$$t_{25,\,0.025} = 2.06 \qquad \text{[Table value]}$$

Since H_1 is two sided the decision rule is reject H_0 if $|t_{25}|$ (calculated value) $\geq t_{25;\,0.05}$ = 2.06.

Here $|t_{25}| = 1.4943143$ (calculated value) $< t_{25;\,0.05} = 2.06$

$\therefore$ We accept H_0 at 5 % level of significance.

Conclusion : From the given results we conclude that average distances required to stop using brakes of two types of cars are equal.

Another method :

p-value = P $[t_{25}|$ > 1.4943 = 0.14769 > level of significance = 0.05 $\therefore$ Accept H_0 at 5% level of significance.

Here we want to find the confidence interval for difference between population means.

(ii) The required 99% confidence interval is given by

$$\left[(\bar{X} - \bar{Y}) - t^* s \sqrt{\frac{1}{n_1} + \frac{1}{n_2}}, \ (x - \bar{y}) + t^* s \sqrt{\frac{1}{n_1} + \frac{1}{n_2}} \right]$$

Here $\alpha = 0.01$, $n_1 = 12$, $n_2 = 15$, $\bar{X} = 7.8$, $\bar{Y} = 9.6$

$$s \sqrt{\frac{1}{n_1} + \frac{1}{n_2}} = 1.20456, \ t^* = t_{n_1 + n_2 - 2; \ \alpha/2} = t_{25; \ 0.005} = 2.787 \text{ (from statistical table)}$$

$\therefore$ The 99% confidence interval for the difference between two population means $\mu_1 - \mu_2$ is

$$[(7.8 - 9.6) - (2.787)(1.20456), (7.8 - 9.6) + (2.787)(1.20456)]$$

$$= \ [-1.8 - 3.3571, -1.8 + 3.3571]$$

$$= \ [-5.1571, 1.5571]$$

Length of C.I. = 2 (3.3571) = 6.7142.

Remark : Our hypothesis to be tested is $H_0 : \mu_1 - \mu_2 = 0$ and

0 lies in interval $[-5.1571, 1.557]$. So we accept H_0 at 1% level of significance.

Use of MS-EXCEL in statistical tests :

For using MS-EXCEL for statistical tests like χ^2 test, t-test, F test first click buttons in sequence $\boxed{\text{Tools}}$ $\rightarrow$ $\boxed{\text{Data Analysis}}$. Let us see how one can solve the problem of testing equality of two population means using MS-EXCEL.

Example 5.4 : The gain in weights (in lbs) of pigs fed on two diets A and B are given below : Test whether the two diets differ significantly regarding their effect on increase in weight.

Diet	Gain in weight
A	25, 32, 30, 34, 24, 14, 32, 24, 30, 31, 35, 25
B	44, 34, 22, 10, 47, 31, 40, 30, 32, 35, 18, 21, 35, 29, 22

Solution : Let μ_1 = Average gain in weight due to diet A

μ_2 = Average gain in weight due to diet B

We want to test $H_0 : \mu_1 = \mu_2$ using MS-EXCEL.

Step 1 : First enter the observations of sample I in one column and sample II in another column of MS-EXCEL sheet as shown below.

	A	B
1	25	44
2	32	34
3	30	22
4	34	10
5	24	47
6	14	31
7	32	40
8	24	30
9	30	32
10	31	35
11	35	18
12	25	21
13		35
14		29
15		22

Fig. 5.5 (a)

Step 2 : Click on Tools and then Data Analysis .

Step 3 : Select t test two sample for means assuming equal variance.

Step 4 : Click on OK then output will be as follows :

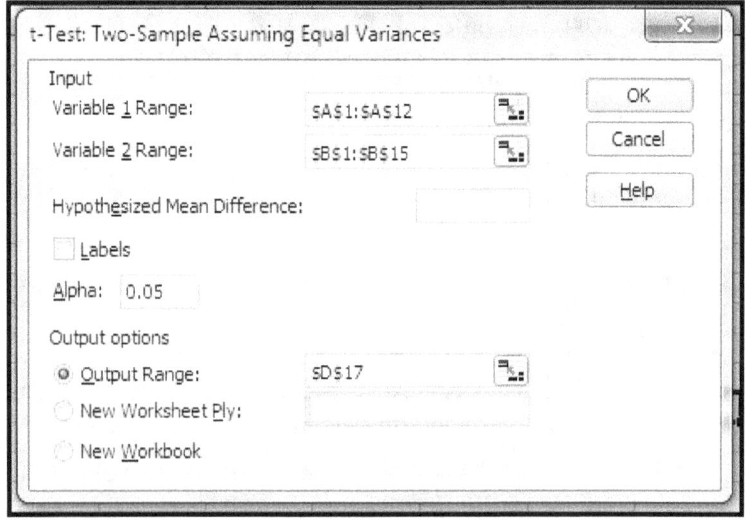

Fig. 5.5 (b)

Step 5 : The following window will appear on the screen.

t-Test: Two-Sample Assuming Equal Variances		
	Variable 1	Variable 2
Mean	28	30
Variance	34.54545455	100.714286
Observations	12	15
Pooled Variance	71.6	
Hypothesized Mean Difference	0	
df	25	
t Stat	-0.610278198	
P(T<=t) one-tail	0.273592836	
t Critical one-tail	1.708140745	
P(T<=t) two-tail	0.547185673	
t Critical two-tail	2.059538536	

Fig. 5.5 (c)

Clearly the output gives the value of t statistic as – 0.771798. t critical one tail value is table value t at 24 d.f. for one tailed test which is 1.710882 on the other hand t critical two tail value is table value t at 24 d.f. for two tailed test. Here it is 2.059538536.

As calculated value < table value, we accept H_0 at 5% level of significance.

Solution using R software :

For testing $H_0 : \mu_1 = \mu_2$ against $H_1 : \mu_1 \neq \mu_2$ we use commands.

```
>x=c (25, 32, 30, 34, 24, 14, 32, 24, 30, 31, 35, 25)
>y=c (44, 34, 22, 10, 47, 31, 40, 30, 32, 35, 18, 21, 35, 29, 22)
> t.test(x,y,var.equal=T)
    Two sample t-test
```

data: x and y

t = – 0.6103, df = 25, p-value = 0.5472

alternative hypothesis : true difference in means is not equal to 0

95 percent confidence interval :

– 8.749507 4.749507

sample estimates :

mean of x mean of y

28 30

From the output since the p-value is greater than level of significance $\alpha = 0.05$ we accept hypothesis.

Example 5.5 : A health club advertised a weight reducing program and claimed that on the average participant in program loses weight in 6 months. A person wanted to verify the claim. The club allowed him to select randomly the records of 10 participants about their weights before and after the program. The data were as follows :

Participant number	1	2	3	4	5	6	7	8	9	10
Weight before joining the club (in lbs)	120	125	115	130	123	119	122	127	128	118
Weight after 6 months from joining the club (in lbs)	111	114	107	120	115	112	112	120	119	112

Do the above data support claim of the health club ?   [Use 1 % level of significance]

Solution : Let, x_i : Weight before joining the club (in lbs).

y_i : Weight after 6 months from joining the club (in lbs).

$d_i : y_i - x_i$ = loss in weight.

Here the situation is suitable for paired t-test

μ_d : average difference in weights.

We want to test $H_0 : \mu_d = 0$ against $H_1 : \mu_d < 0$. The test statistic is

$$t_{n-1} = \frac{\overline{d}}{\dfrac{s}{\sqrt{n}}}$$

where, $\overline{d} = \dfrac{\Sigma d_i}{n}$ and $s^2 = \dfrac{\Sigma d_i^2 - n\,\overline{d}^2}{n-1}$

Here,

d_i	−9	−11	−8	−10	−8	−7	−10	−7	−9	−6	$\Sigma d_i = -85$

$$\Sigma d_i^2 = 745,$$

$$\overline{d} = -8.5$$

$$s^2 = \frac{745 - (10)\,(-8.5)^2}{9} = \frac{22.5}{9} = 2.5$$

$$\therefore \quad s = 1.5811388$$

$$\therefore \quad t_9 = \frac{-8.5}{\dfrac{1.5811388}{\sqrt{10}}} = -16.9999 \ \text{[Calculated value]}$$

$$t_{9;\,0.01} = -2.821 \qquad\qquad \text{[Table value]}$$

[Since $H_1 : \mu_d < 0$ is one sided see the value in table corresponding to n = 9 and probability 0.02]

$$\therefore \qquad t_9 = -16.999 < t_{9;\,0.01} = -2.821$$

We reject H_0 at 1% level of significance.

Another method : $\therefore$ p-value is P $[t_9 < -16.9999]$ = P $[t_9 > 16.9999]$ = 3.2×10^{-8} < level of significance = 0.01.

$\therefore$ Reject H_0.

Conclusion : The given data support the claim of the health club that on the average, participant in the program loses weight during 6 months.

Paired t-test using MS-EXCEL : The steps involved to carryout paired t test using MS-EXCEL are identical to that in case of two sample t test for means.

Step 1 : Enter the observations of two samples on two different $\boxed{\text{Tools}}$ and then select the $\boxed{\text{Data Analysis}}$ Tool.

Step 2 : Click on $\boxed{\text{Tools}}$ and then $\boxed{\text{Data Analysis}}$.

Step 3 : Select $\boxed{\text{t-test paired two sample for means}}$.

Step 4 : The following window will appear on screen.

H	I
120	111
125	114
115	107
130	120
123	115
119	112
122	112
127	120
128	119
118	112

Fig. 5.6 (a)

Step 5 : Give the range for variable 1 and variable 2 as shown in following Fig. 5.6 (b).

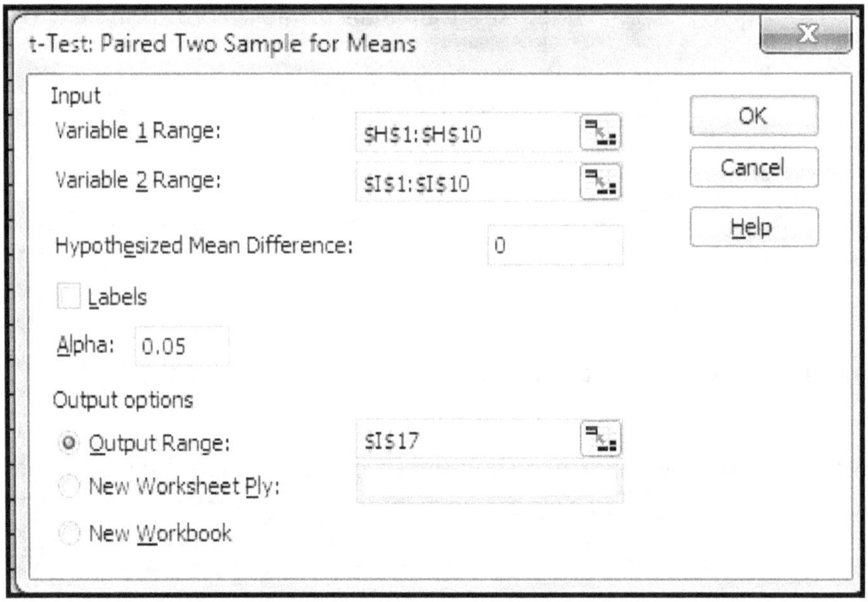

Fig. 5.6 (b)

Step 6 : Click on OK and get output usual form for t text.

t-Test: Paired Two Sample for Means		
	Variable 1	Variable 2
Mean	122.7	114.2
Variance	23.12222222	18.62222222
Observations	10	10
Pearson Correlation	0.945622184	
Hypothesized Mean Difference	0	
df	9	
t Stat	17	
P(T<=t) one-tail	1.89367E-08	
t Critical one-tail	1.833112923	
P(T<=t) two-tail	3.78734E-08	
t Critical two-tail	2.262157158	

Fig. 5.6 (c)

We compare t stat (which is calculated) with critical one sided (which t table value at 5% level of significance) since H_1 is one sided. Here t stat = 17 > t critical one sidel = 1.833113, we reject H_0.

Solution using R software :

For testing $H_0 : \mu_d = 0$ against $H_1 : \mu_d \neq 0$ we use commands

```
>x=c (120, 125, 115, 130, 123, 119, 122, 127, 128, 118)
>y=c (111, 1114, 107, 120, 115, 112, 112, 120, 119, 112)
>t.test(y,x,paired=T,alt="1")
```

Paired t-test

data: y and x

t = – 17, df = 9, p-value = 1.894e – 08

alternative hypothesis : true difference in means is less than 0

95 percent confidence interval :

– Inf – 7.583444

sample estimates :

mean of the difference

– 8.5

5.2 Tests Based on χ^2 Distribution

[A] Test of population variance σ^2

Let us consume that the observations in the population under consideration have $N(\mu, \sigma^2)$ distribution. We wish to test $H_0 : \sigma^2 = \sigma_0^2$ against $H_1 : \sigma^2 \neq \sigma_0^2$ where σ_0^2 is the specified value of σ^2. The population mean μ may be known or unknown. We consider both the cases as follows :

Case (i) : Testing $H_0 : \sigma^2 = \sigma_0^2$ when μ is known :

Let $x_1, x_2 \ldots x_i, \ldots x_n$ be a random sample drawn from the parent normal population.

$\because x_i \to$ i.i.d. $N(\mu, \sigma^2);\ i = 1, 2 \ldots n$

$$\frac{x_i - \mu}{\sigma} \to \text{i.i.d. } N(0, 1);\ i = 1, 2 \ldots n$$

$$\therefore\ \frac{\sum_{i=1}^{n} (x_i - \mu)^2}{\sigma^2} \to \chi^2\text{-distribution with } n \text{ d.f.}$$

Hence on $H_0 : \sigma^2 = \sigma_0^2$ the statistic

$$\frac{\sum_{i=1}^{n} (x_i - \mu)^2}{\sigma_0^2} \to \chi_n^2$$

The rejection region in this case is shaded region as shown in figure 5.4.

Thus we reject H_0 at level of significance α if

$$\chi_n^2 = \frac{\sum(x_i - \mu)^2}{\sigma_0^2} \geq \chi_{n,\frac{\alpha}{2}}^2 \text{ or } \chi_n^2 \leq \chi_{n,\,1-\alpha/2}^2 \text{ and accept } H_0 \text{ otherwise.}$$

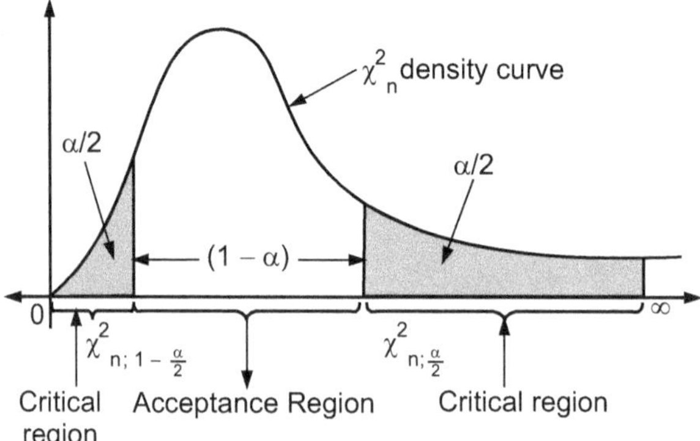

Fig. 5.7

Note :

1. χ^2-distribution is not symmetric. Hence, equal tail criterion is used. It means area of each tail is taken as $\alpha/2$.

2.   The value of a statistic in χ^2 test is always positive.

3.   If $H_1 : \sigma^2 > \sigma_0^2$ then rejection region at l.o.s. is $\chi_n^2 \geq \chi_{n;\ \alpha}^2$ which is shaded region as shown in Fig. 5.9. On the other hand when $H_1 : \sigma^2 < \sigma_0^2$ the critical region at level of significance α is $\chi_n^2 \leq \chi_{n;\ 1-\alpha}^2$ as shown in Fig. 5.10.

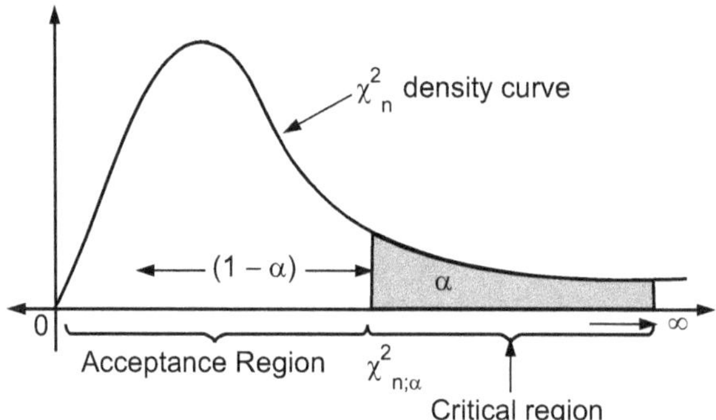

Fig. 5.8

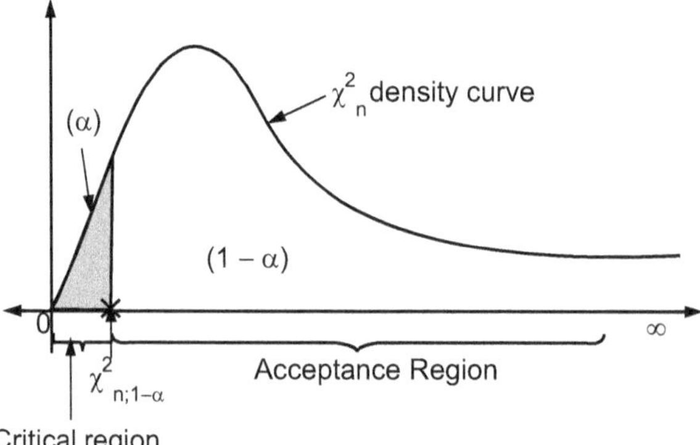

Fig. 5.9

Case (ii) : Testing $H_0 : \sigma^2 = \sigma_0^2$ when μ is unknown : The statistic

$$\chi^2 = \frac{\sum\limits_{i=1}^{n} (X_i - \bar{X})^2}{\sigma_0^2}$$

has χ^2-distribution with $(n-1)$ degrees of freedom (d.f.)

Hence, for testing $H_0 : \sigma^2 = \sigma_0^2$ against $H_1 : \sigma^2 \neq \sigma_0^2$ the critical region at *l.o.s.* α if $\chi_{n-1}^2 \geq \chi_{n-1;\frac{\alpha}{2}}^2$ or $\chi_{n-1}^2 \leq \chi_{n-1;\,1-\alpha/2}^2$.

Note : The rejection region for testing $H_0 : \sigma^2 = \sigma_0^2$ l.o.s. α is $\chi_{n-1}^2 \geq \chi_{n-1;\,\alpha}^2$ for $H_1 : \sigma^2 > \sigma_0^2$ and it is $\chi_{n-1}^2 \leq \chi_{n-1;\,1-\alpha}^2$ if $H_1 : \sigma^2 < \sigma_0^2$ has χ^2.

[B] Test for Goodness of Fit

For a given data (frequency distribution) we try to fit some probability distribution. Since there are several probability distributions, which distribution will fit properly may be a question of interest. In such cases we want to test the appropriateness of the fit. Hence, we desire to test H_0 : Fitting of the probability distribution to given data is proper (good). The test based on χ^2-distribution is used to test this H_0. It is called χ^2-test of goodness of fit.

In this case we compare the expected and observed frequencies. Thus we can take H_0 : There is no significant difference between observed and (theoretical) expected frequencies. The test is carried out as follows :

Suppose $o_1, o_2, \dots o_i, \dots, o_k$ be a set of observed frequencies and $e_1, e_2, \dots e_i, \dots, e_k$ be corresponding expected frequencies obtained under H_0.

$$\sum_{i=1}^{k} o_i = N = \sum_{i=1}^{k} e_i$$

p = Number of parameters estimated for fitting the probability distribution.

If H_0 is true, then the statistic.

$$\chi^2 = \sum_{i=1}^{k} \frac{(o_i - e_i)^2}{e_i} = \sum_{i=1}^{k} \left(\frac{o_i^2}{e_i}\right) - N$$

has χ^2-distribution with $(k - p - 1)$ degrees of freedom. In this case the critical region at l.o.s. α is,
$$\chi_{k-p-1}^2 \geq \chi_{k-p-1;\,\alpha}^2$$

It is shown by the shaded region in Figure 5.11.

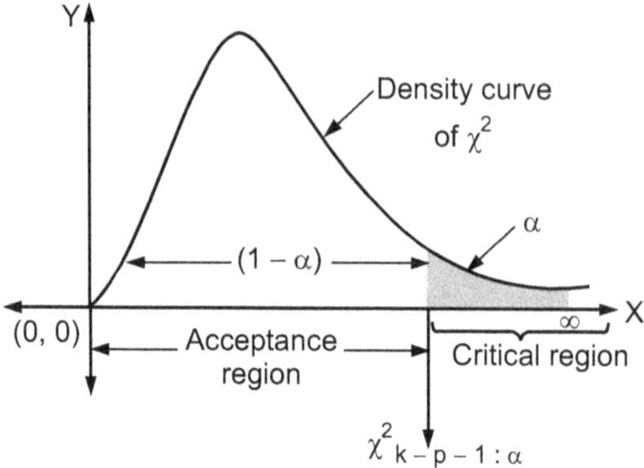

Fig. 5.10

Thus we reject H_0 at l.o.s. α if,

$$\chi^2_{k - p - 1} \geq \chi^2_{k - p - 1; \alpha}$$

Note :

1. We can apply this test if expected frequencies are greater than or equal to 5 (i.e. $e_i \geq 5$) and total of cell frequencies is sufficiently large (greater than 50).

2. When expected frequency of a class is less than 5, the class is merged into neighbouring class along with its observed and expected frequencies until total of expected frequencies becomes ≥ 5. This procedure is called 'pooling of the classes'. In this case k is the number of class frequencies after pooling.

3. It is obvious that if any parameters is not estimated while fitting a probability distribution or obtaining expected frequencies the value of p is zero.

4. This test is not applicable for testing goodness of fitting of straight line or curves such as second degree curve, exponential curve etc.

[C]   Test for Independence of Two Attributes :

Suppose that the given data are classified into r levels of attribute A denoted by $A_1, A_2, ..., A_i, ..., A_r$ and s levels of attribute B represented by $B_1, B_2 ..., B_j, ..., B_s$. Then different class frequencies [Cell frequencies] can be represented in the following tabular form.

A \ B	B_1	B_2	...	B_j	...	B_s	Total
A_1	O_{11}	O_{12}	...	O_{1j}	...	O_{1s}	(A_1)
A_2	O_{21}	O_{22}	...	O_{2j}	...	O_{2s}	(A_2)
$\vdots$							$\vdots$
A_i	O_{i1}	O_{i2}	...	O_{ij}	...	O_{is}	(A_i)
$\vdots$							$\vdots$
A_r	O_{r1}	O_{r2}	...	O_{rj}	...	O_{rs}	(A_r)
Total	(B_1)	(B_2)	...	(B_j)	...	(B_s)	N

The above type of table containing r rows and s columns is called $r \times s$ contingency table. O_{ij} is called observed frequency corresponding to $(i, j)^{th}$ cell $i = 1, 2, ..., r, j = 1, 2, ..., s$.

$$N = \sum_{i=1}^{r} \sum_{j=1}^{s} O_{ij} = \text{Total observed frequency,}$$

$$(A_i) = \sum_{j=1}^{s} O_{ij} = \text{Total of observed frequencies in the } i^{th} \text{ row, } i = 1, 2 ... r$$

$$(B_j) = \sum_{j=1}^{r} O_{ij} = \text{Total of observed frequencies in the } j^{th} \text{ column, } j = 1, 2 ... s$$

We wish to test H_0 : Two attributes A and B are independent against H_1 : The attributes A and B are not independent (associated or disassociated to each other). Under the hypothesis of independence of attributes (i.e. H_0), the expected frequencies corresponding the given observed frequencies are obtained as;

$$e_{ij} = \frac{(A_i)(B_j)}{N} ; i = 1, 2 ... r, j = 1, 2 ... s$$

$$= \frac{(\text{Total of observed frequency in the } i^{th} \text{ row}) \times (\text{Total of observed frequencies in the } j^{th} \text{ column})}{\text{Grand total of all observed frequencies}}$$

where e_{ij} : expected frequency of $(i, j)^{th}$ cell. Hence, using the above formula we can find all expected frequencies. Then table of expected frequencies can be prepared as follows :

A \ B	B_1	B_2	...	B_j	...	B_s
A_1	e_{11}	e_{12}	...	e_{ij}	...	e_{1s}
A_2	e_{21}	e_{22}	...	e_{2j}	...	e_{2s}
$\vdots$						
A_i	e_{i1}	e_{i2}	...	e_{ij}	...	e_{is}
$\vdots$						
A_r	e_{r1}	e_{r2}	...	e_{rj}	...	e_{rs}

$$\text{Total} \qquad \sum \sum e_{ij} = N$$

Then if H_0 is true, the statistic

$$\chi^2 = \sum_{i=1}^{r} \sum_{j=1}^{s} \frac{(O_{ij} - e_{ij})^2}{e_{ij}} \qquad \text{... (i)}$$

$$= \sum_{i=1}^{r} \sum_{j=i}^{s} \left(\frac{O_{ij}^2}{e_{ij}} \right) - N$$

follows χ^2-distribution with $(r - 1)(s - 1)$ d.f. The critical region at $\alpha\%$ level of significance is the shaded region shown below :

Thus we reject H_0 at $\alpha\%$ level of significance if $\chi^2_{(r-1)(s-1)} \geq \chi^2_{(r-1)(s-1);\,\alpha}$ and accept H_0 otherwise.

Note :

1. Here critical region is always right tailed.

2. In particular if we have $r = 2$, $s = 2$ i.e. two attributes A and B are at two levels each, then the contingency table is

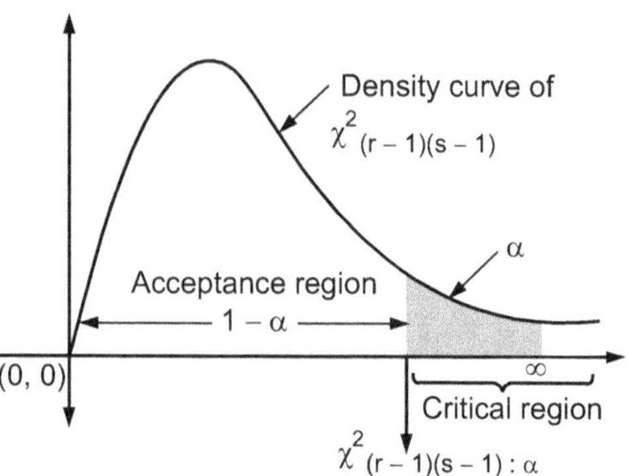

Fig. 5.11

A B	B_1	B_2	Total
A_1	a	b	a + b
A_2	c	d	c + d
Total	a + c b + d		N = a + b + c + d

In this case H_0 : Attributes A and B are independent, the test statistic given by (1) simplifies to

$$\chi^2 = \frac{N(ad - bc)^2}{(a + b)(c + d)(a + c)(b + d)}$$

which follows χ^2-distribution with 1 d.f.

Justification : We know that the test statistic for testing such H_0 for $r \times s$ contingency table is given by,

$$\chi^2_{(r-1)(s-1)} = \sum_i \sum_j \frac{(O_{ij} - e_{ij})^2}{e_{ij}} \qquad \ldots (i)$$

In this case $O_{11} = a$, $O_{12} = b$, $O_{21} = c$, $O_{22} = d$

Under H_0 : Attributes A and B are independent, the expected frequencies will be as follows :

$$e_{11} = \frac{(a + b)(a + c)}{N},$$

$$e_{12} = \frac{(a + b)(b + d)}{N}$$

$$e_{21} = \frac{(c + d)(a + c)}{N}$$

and $\qquad e_{22} = \frac{(b + d)(c + d)}{N}$

Putting r = 2, s = 2 and the values of observed as well as expected frequencies in (i), we get

$$\chi_1^2 = \frac{(O_{11} - e_{11})^2}{e_{11}} + \frac{(O_{12} - e_{12})^2}{e_{12}} + \frac{(O_{21} - e_{21})^2}{e_{21}} + \frac{(O_{22} - e_{22})^2}{e_{22}}$$

$$\text{First term on R.H.S.} = \frac{(o_{11} - e_{11})^2}{e_{11}}$$

$$= \frac{\left[a - \dfrac{(a + b)(a + c)}{a + b + c + d}\right]^2}{\dfrac{(a + b)(a + c)}{a + b + c + d}}$$

$$= \frac{[a(a + b + c + d) - (a + b)(a + c)]^2}{(a + b + c + d)^2} \cdot \frac{(a + b + c + d)}{(a + b)(a + c)}$$

$$= \frac{(ad - bc)^2}{N(a + b)(a + c)}$$

Other terms on R.H.S. can be written in similar way

$$\therefore \ \chi_1^2 = \frac{(ad - bc)^2}{N(a + b)(a + c)} + \frac{(ad - bc)^2}{N(a + b)(b + d)} + \frac{(ad - bc)^2}{N(c + d)(a + c)} + \frac{(ad - bc)^2}{N(B + d)(c + d)}$$

$$= \frac{[(c + d)(b + d) + (a + c)(c + d) + (a + b)(b + d) + (a + b)(a + c)](ad - bc)^2}{N(a + b)(c + d)(a + c)(b + d)}$$

$$\therefore \ \chi_1^2 = \frac{N(ad - bc)^2}{(a + b)(c + d)(a + c)(b + d)}$$

Note :

Yate's Correction : If in a 2×2 contingency table any cell frequency is less than 5 then the test statistic χ^2 is corrected in a specific way. This correction is due to Yate's and hence is known as Yate's correction.

It can been shown that when any cell frequency in the contingency table of order (2×2) is less than 5, the continuity of χ^2 distribution curve is not maintained. Yates suggested a correction to remove this drawback. It is called Yate's correction. The modified formula obtained by him in this case is as follows :

$$\chi_1^2 = \frac{\left\{|ad - bc| - \dfrac{N}{2}\right\}^2 N}{(a + b)(c + d)(a + c)(b + d)}$$

Hence, whenever a cell frequency is less than 5, Yate's corrected formula given above should be used.

5.6 McNemar's Test :

We have seen the test of independence of two attributes each at 2 levels (2×2 contingency table). However this test cannot be used to test the hypothesis of symmetry for 2×2 contingency table. In this case McNemar's test is applied we want to test that the two attributes are equally effective. Suppose the 2×2 contingency table is given by

B\A	B_1	B_2	Total
A_1	O_{11}	O_{12}	$O_{1\cdot}$
A_2	O_{21}	O_{22}	$O_{2\cdot}$
Total	$O_{\cdot 1}$	$O_{\cdot 2}$	N

where O_{ij}'s are observed frequencies $i, j = 1, 2$. The corresponding table of expected frequencies is given by

B\A	B_1	B_2	Total
A_1	e_{11}	e_{12}	$e_{1\cdot}$
A_2	e_{21}	e_{22}	$e_{2\cdot}$
Total	$e_{\cdot 1}$	$e_{\cdot 2}$	N

Accordingly the table of probabilities is

B\A	B_1	B_2	Total
A_1	$\dfrac{e_{11}}{N} = P_{11}$	$\dfrac{e_{12}}{N} = P_{12}$	$\dfrac{e_{1\cdot}}{N} = P_1$
A_2	$\dfrac{e_{21}}{N} = P_{21}$	$\dfrac{e_{22}}{N} = P_{22}$	$\dfrac{e_{2\cdot}}{N} = P_{2\cdot}$
Total	$\dfrac{e_{\cdot 1}}{N} = P_{\cdot 1}$	$\dfrac{e_{\cdot 2}}{N} = P_{\cdot 2}$	1

In this test our null hypothesis H_0 is of equality of marginal distributions against **two sided** alternative that they are not equal.

In other words :

$$H_0 \ : \ P_{1\cdot} = P_{\cdot 1} \text{ against } H_1 : P_{1\cdot} \neq P_{\cdot 1}$$

OR $\qquad H_0 \ : \ P_{2\cdot} = P_{\cdot 2} \text{ against } H_1 : P_{\cdot 2} \neq P_{\cdot 2}$

In this case under H_0, the test statistic $\dfrac{(O_{12} - O_{21})^2}{O_{12} + O_{21}}$ follows χ^2 distribution with 1 d.f.

This test is called MC Neymar's test. The value of this test statistic is compared with the corresponding table value for given level is significance.

Hence we reject H_0 at *l.o.s.* α if $\chi_1^2 \geq \chi_{1;\,\alpha}^2$ (Table value at *l.o.s.* α) and accept H_0 otherwise.

If H_0 is rejected we conclude that the effects for two attributes are not equal.

Remarks :

1. The above test statistic is used for two sided alternative only.

The null hypothesis in this test is identical to that of equality of population proportions.

2. The null hypothesis looks like equality of two population means but here we do not consider two **different** populations.

Solved Examples

Example 5.6 : In a department examination, the candidates of both sexes yielded as presented in following table :

Sex	Pass	Fail
Male	1	6
Female	7	6

Can it be inferred that the result of the test is related to the sex of the candidates ?

Solution : Let   A :   Result of the test

B :   Sex of the candidate

H_0 :   Attributes A and B are independent

$\because$ The first row, first columne cell frequency is less than 5, we have to apply Yate's correction.

$$\chi_1^2 \;=\; \frac{\{|ad - bc| - N/2\}^2\, N}{(a + b)\,(c + d)\,(a + c)\,(b + d)} \qquad \text{Here } a = 1, b = 6, c = 7, d = 6$$

$$=\; \frac{\{|1 \times 6 - 6 \times 7| - 20/2\}^2\, 20}{6 \times 13 \times 8 \times 12}$$

$$=\; \frac{(36 - 10)^2\, 20}{6 \times 13 \times 8 \times 12} = \frac{13520}{8736} = 1.55 \text{ [calculated values]}$$

$$\chi_{1,\,0.05}^2 \;=\; 3.841 \text{ [Table value]}$$

We accept H_0 at 5% level of significance.

Conclusion : The result of the test is not related to sex of the candidate.

Example 5.7 : Various investigations were made for testing the incidence of heavy infection of material parasite plasmodium falciparum in groups of children with heterozygotes and group of children with the normal homozygotes. The findings in the investigation were as follows :

Groups of children	Heavy infections	Non-infected
Children with heterozygotes	18	50
Children with normal homozygotes	62	127

Test whether the heterozygotes are better protected than normal homozygotes from malarial infections.

Solution :

Let,   A : Infection of material parasite.

B : Children with heterozygotes.

We want to test H_0 : Attributes A and B are independent.

We are given 2×2 contingency table with $a = 18$, $b = 50$, $c = 62$, $d = 127$. Hence the test statistic is

$$\chi_1^2 = \frac{(ad - bc)^2 \, N}{(a + b)(c + d)(a + c)(b + d)}$$

$$= \frac{[(18)(127) - (62)(50)]^2 \, 257}{(68)(189)(80)(177)}$$

$$= 0.9357241 \qquad \text{[Calculated value]}$$

$$\chi_{1;\,0.05}^2 = 3.841$$

$$\therefore \quad \chi_1^2 = 0.9357241 < \chi_{1;\,0.05}^2 = 3.841 \qquad \text{[Table value]}$$

we accept H_0 at 5 % level of significance.

Second approach : p-value $= P\,[\chi_1^2 > 0.9357] = 0.33886 >$ level of significance. $= 0.05$ $\therefore$ Accept H_0.

Conclusion : Attributes A and B are independent. Thus the heterozygotes may not be better protected than normal homozygotes from material infections.

Example 5.8 : A sample 390 males was selected to find association between their occupation and occupation of their fathers. It revealed following information.

Son's occupation Father's occupation	Agriculture	Business	Service	Other
Agriculture	33	18	25	19
Business	15	29	26	30
Service	15	16	38	16
Other	17	25	31	37

Test whether the son's occupation is independent of father's occupation (use 5 % level of significance)

Solution : Here we want to test H_0 : Two attributes son's occupation (A) and father's occupation (B) are independent of each other against H_1 : They are associated with each other.

The given frequencies are observed frequencies (O_{ij}). The corresponding expected frequencies e_{ij} are obtained using formula

$$e_{ij} = \frac{(A_i)(B_j)}{N} = \frac{(A_i)(B_j)}{390} \text{ for } i, j = 1, 2, 3, 4$$

Note : e_{ij}'s are not to be rounded-off to the integers.

Table of observed frequencies (o_{ij})

	B_1	B_2	B_3	B_4	Total
A_1	33	18	25	19	95
A_2	15	29	26	30	100
A_3	15	16	38	16	85
A_4	17	25	31	37	110
Total	80	88	120	102	N = 390

Table of expected frequencies (e_{ij})

19.487179	21.435897	29.230769	24.846153
20.51282	22.564102	30.76923	26.153846
17.435897	19.179487	26.153846	22.230769
22.564102	21.820512	33.846153	28.76923

Table of $\dfrac{O_{ij}^2}{e_{ij}}$

55.8829	15.1148	21.3816	14.5294
10.9688	37.2716	21.9700	34.4118
12.9044	13.3476	55.2118	11.5156
12.8080	25.1808	28.3932	47.5856

The test statistic is

$$\chi^2_{(r-1)(s-1)} = \chi^2_{3 \times 3}$$

$$= \chi^2_9 = \underset{i}{\Sigma}\,\underset{j}{\Sigma}\left(\frac{O_{ij}^2}{e_{ij}}\right) - N$$

$$= 418.4779 - 390$$

$$\therefore \quad \chi^2_9 = 28.4779 \qquad \text{[Calculated value]}$$

$$\chi^2_{9;\,0.05} = 16.919 \qquad \text{[Table value]}$$

$$\because \chi^2_9 \text{ (calculated value)} = 28.4779 > \chi^2_{9;\,001} = 16.919 \qquad \text{[Table value]}$$

reject H_0 at 5 % level of significance.

Alternatively p-value $= P\,[\chi^2_9 > 28.4779] = 0.00079 <$ level of significance $= 0.05$.

$\therefore$ Reject H_0.

Chi-square test using MS-EXCEL :

Step 1 : Enter observed frequencies in table ($B_2 : E_5$).

Step 2 : Compute expected frequencies using MS-EXCEL sheet separately are enter in table ($B_9 : B_{12}$).

Step 3 : Click on functions $\boxed{f_x} \longrightarrow \boxed{\text{CHITEST}} \rightarrow \boxed{\text{OK}}$.

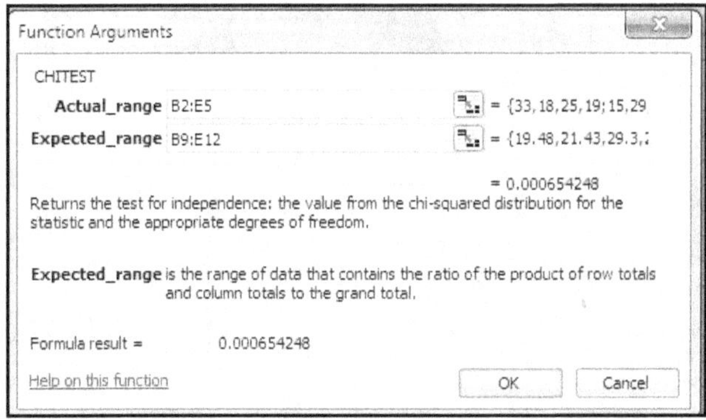

Fig. 5.12 (a)

Fig. 5.12 (b)

Step 4 : Select actual range for observed frequency.

Select actual range for expected frequency.

Click OK .

Fig. 5.12 (c)

Formula result gives P value = 0.00065.

If P value < 0.05 reject of 5% level of significance.

Conclusion : Occupation of son may depend on occupation of father.

Solution using R software :

```
>x=c (33,18,25,19, 15, 29, 26, 30, 15, 16, 38, 16, 17, 25, 31, 37)
>m=matrix(x,byrow=T,ncol=4)
>chisq.test(m)
```

Pearson's Chi-square test

data: m

X-squared = 28.4776, df = 9, p-value = 0.0007934

From the output since the p-value is less than level of significance α = 0.05 we reject hypothesis.

Example 5.9 : A sample of 200 patients is drawn to test whether the audio quality of right ear is same as that of left ear. (Use 5% *l.o.s.*)

The data obtained are as follows :

		Audio quality of left ear	
		Satisfactory	Not satisfactory
Audio quality of right ear	Satisfactory	160	17
	Unsatisfactory	15	08

Solution : We define attribute

$$A \; : \; \text{Quality of right ear}$$
$$B \; : \; \text{Quality of left ear}$$

Here $\qquad H_0 \; : \;$ The audio qualities of left eye and right eye aresame

against $H_1 \; : \;$ The audio qualities of right and left ears a different

Given : $O_{11} = 160$, $O_{12} = 17$, $O_{21} = 15$, $O_{22} = 08$. We shall apply McNeymar's test.

$$\text{Test statistic is } \chi_1^2 \; = \; \frac{(O_{12} - O_{21})^2}{O_{12} + O_{21}} = \frac{(17 - 15)^2}{17 + 15} = \frac{4}{32} = \frac{1}{8} = 0.125 \text{ [Calculated value]}$$

$$\chi_{1;\, 0.05}^2 \; = 3.841 \qquad \text{[Table value]}$$

Decision : Accept H_0 at 5% level of significance.

Conclusion : The audio quality of right ear may be same as that of left ear.

Example 5.10 : A manager of a coal mining factory claimed that the standard deviation of the percentage of ash content in coal supplied by the factory is 2.1. A sample of 12 bags was drawn to justify the claim. They shared the percentage of ash contents as 11.6, 12.9, 14.3, 16.2, 14.2, 13.2, 15.0, 15.2, 13, 12.6, 13.5, 11.9.

(i)　　Test normality using normal probability paper.

(ii)　　Does the sample support the claim of manager of the factory ?

Solution : Let σ^2 be variance of the percentage of ash content in coal supplied by the company.

$X_{(i)}$	i	$P = \dfrac{i - 0.5}{n} = \dfrac{i - 0.5}{12}$	100 P
11.60	1	0.0417	4.17
11.90	2	0.1250	12.50
12.60	3	0.2083	20.83
12.70	4	0.2917	29.17
13.00	5	0.3750	37.50
13.20	6	0.4583	45.83
13.50	7	0.5417	54.17
14.20	8	0.6250	62.50
14.30	9	0.7083	70.83
15.00	10	0.7917	79.17
15.20	11	0.8750	87.50
16.20	12	0.9583	95.83

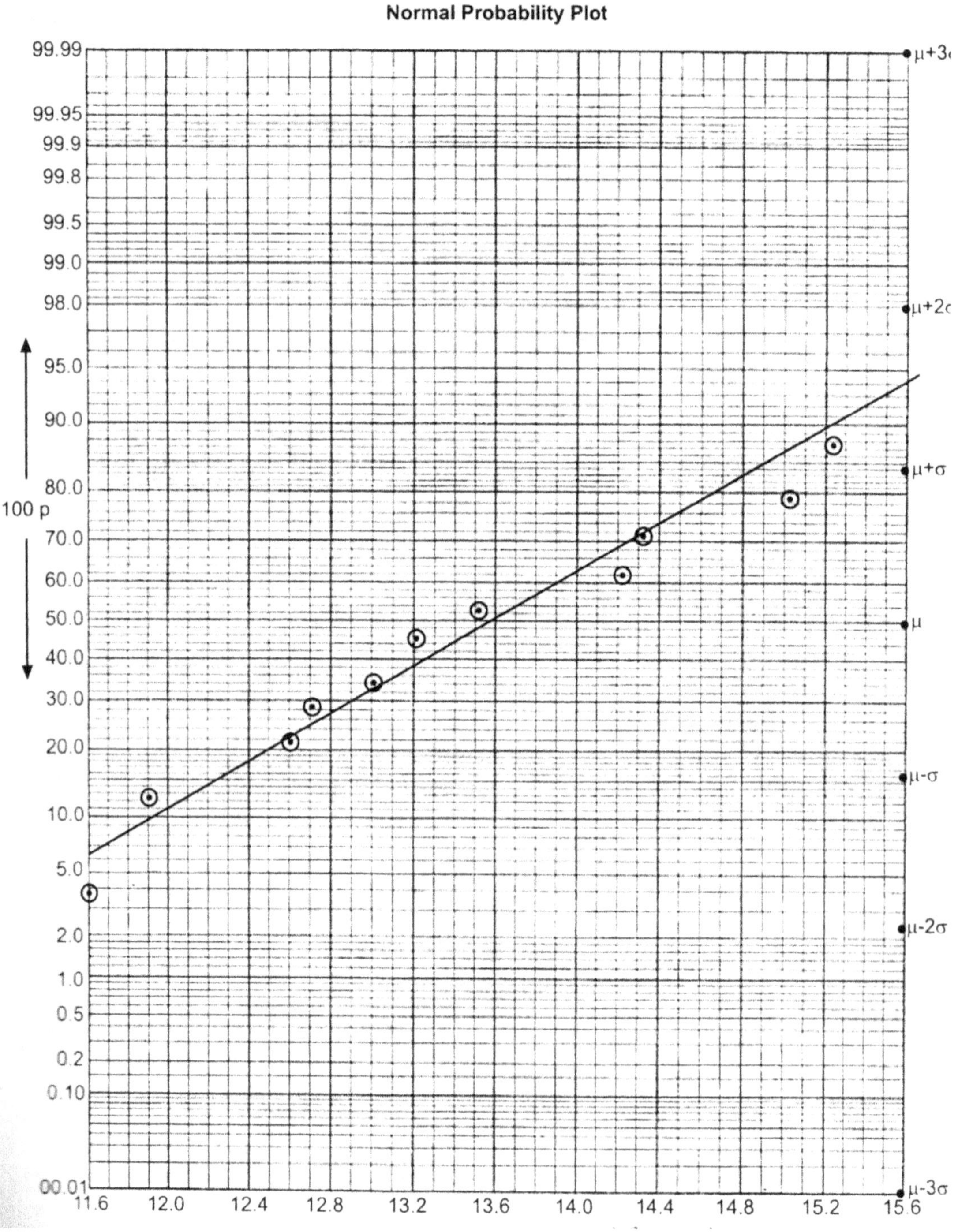

Normal Probability Plot

Fig. 5.13

$X_{(i)}$

From the graph the observations seems to come from normal distribution.

We want to test $H_0 : \sigma^2 = \sigma_0^2 = (2.1)^2$ against $H_1 : \sigma^2 \neq (2.1)^2$. The test statistic is

$$\chi_{n-1}^2 = \frac{\Sigma (x_i - \bar{x})^2}{\sigma_0^2}$$

$$= \frac{\Sigma x_i^2 - n\,\bar{x}^2}{\sigma_0^2} \qquad [\because \text{Population mean is unknown}]$$

Here,

$$\bar{X} = \frac{\Sigma X_i}{n} = \frac{163.6}{12}$$

$$= 13.6333$$

$$\Sigma X_i^2 = 2251.44$$

$$\therefore \quad \chi_{11}^2 = \frac{2251.44 - 2230.4023}{4.41}$$

$$= \frac{21.0377}{4.41}$$

$$\therefore \quad \chi_{11}^2 = 4.7704535 \qquad \text{[Calculated value]}$$

$$< \chi_{11;\,0.05}^2 = 19.675 \qquad \text{[Table value]}$$

$\therefore$ We accept H_0 at 5 % l.o.s.

Second method : p-value = $P\,[\chi_{11}^2 > 4.77] = 0.9418 > 0.05$ $\therefore$ Accept H_0 at 5% level of significance.

Conclusion : The sample supports the claim of manager of the factory that the standard deviation of percentage of ash content in coal is 2.1.

Example 5.11 : The machine producing ball bearing is operating properly if the variance of the diameter is 1.44 $(cm)^2$. A random sample of 14 ball bearings showed variance of 2.08 $(cm)^2$. Will you conclude that the machine is operating properly based on these data ?

Solution : In this problem we shall test

$$H_0 : \sigma^2 = \sigma_0^2 = 1.44 \text{ against } H_1 : \sigma^2 > \sigma_0^2$$

$\therefore$ The statistic is given by,

$$\chi_{n-1}^2 = \frac{\displaystyle\sum_{i=1}^{n} (x_i - \bar{x})^2}{\sigma_0^2}$$

$$\text{Sample variance} = \frac{\Sigma (x_i - \bar{x})^2}{n} = 2.08 \qquad \text{[Given]}$$

$$\therefore \quad \sum_{i=1}^{n} (x_i - \bar{x})^2 = (14)(2.08) = 29.12$$

$$\therefore \quad \chi^2_{14} = \frac{29.12}{1.44}$$

$$= 20.22222 \quad \text{[Calculated value]}$$

$$\chi^2_{14;\, 0.05} = 23.685 \qquad \text{[Table value]}$$

Since H_1 is one sided, the decision rule is reject H_0 if χ^2_{14} calculated value) $> \chi^2_{14;\, 0.05}$ and accept H_0 otherwise.

Hence we accept H_0 at 5 % level of significance.

Another approach : p-value : P $[\chi^2_{14} > 20.22222] = 0.8305 >$ level of significance $= 0.05$
$\therefore$ Accept H_0.

Conclusion : Based on these data we conclude that the machine is operating properly.

Example 5.12 : A nationalized bank utilizes four teller windows to render fast service to the customers. On a particular day 800 customers were observed. They were given service at the different windows as follows :

Window number	Number of customers
1	150
2	250
3	170
4	230

Test whether the customers are uniformly distributed over the windows.

Solution : Here, we want to test H_0 : Customers are uniformly distributed over the windows. i.e. H_0 : customers on all windows are equal against H_1 : They are not equal on all windows.

Under H_0, the expected frequencies are :

Window number	Expected number of customers (e_i)
1	200
2	200
3	200
4	200

The test statistic is

$$\chi^2_{k-p-1} = \sum_{i=1}^{k} \frac{(O_i - e_i)^2}{e_i}$$

Here number of parameters estimated $= p = 0$, $k = 4$.

$$\therefore \qquad \chi^2_3 = 34 \qquad\qquad \text{[Calculated value]}$$

$$> \chi^2_{3;\,0.05} = 7.815 \qquad\qquad \text{[Table value]}$$

We reject H_0 at 5 % $l.o.s.$

Also $P\,[\chi^2_3 > 34] = 0.00001 = $ p-value $<$ level of significance $= 0.05$ $\therefore$ Reject H_0.

Conclusion : The customers in the nationalized bank may not be uniformly distributed over different windows.

Note : Chi-square test of goodness of fit makes a comparison between observed frequencies (O_i) and expected frequencies (e_i). We can visualise the difference in O_i and e_i by multiple bar diagram as follows.

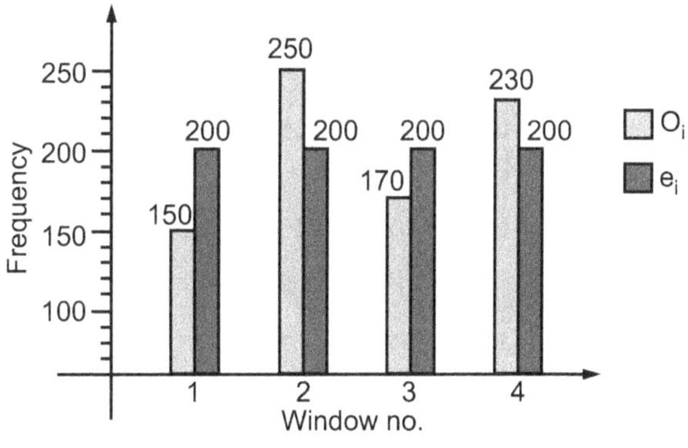

Fig. 5.14

Example 5.13 : Among 64 off-springs of a certain cross between guinea pigs 34 were red, 10 were black and 20 were white. According to a genetic model these numbers should be in the ratio 9 : 3 : 4. Are the data consistent with the model at 5 % level ?

Solution : Here H_0 : The offsprings red, black and white in colour are in the ratio 9 : 3 : 4.

In this problem, $N = 64$. Hence observed and expected frequencies are as follows :

Type	Red	Black	White
Observed frequencies (o_i)	34	10	20
Expected frequencies (e_i)	$\frac{9}{16} \times 64 = 36$	$\frac{3}{16} \times 64 = 12$	$\frac{4}{16} \times 54 = 16$

To test H_0, the test statistic is

$$\chi^2_{k-p-1} = \sum_{i=1}^{k} \frac{(O_i - e_i)^2}{e_i} \qquad \text{here } p = 0 \text{ and } k = 3$$

$$\therefore \qquad \chi^2_2 = 1.444444 \qquad \text{[Calculated value]}$$

$$\chi^2_{2;\,0.05} = 5.991 \qquad \text{[Table value]}$$

$$\therefore \qquad \chi^2_2 = 1.444444 \text{ (c.v.)}$$

$$< \chi^2_{2;\,0.05} = 5.99 \qquad \text{[Table value]}$$

We accept H_0 at 5 % level of significance.

Another method : p-value = P $[\chi^2_2 > 1.444] \leq 0.45 > 0.005$ $\therefore$ Accept H_0 at 5% l.o.s.

Conclusion : The data are consistent with the genetic model that the offspring red, black and white in colour are in the ratio 9 : 3 : 4.

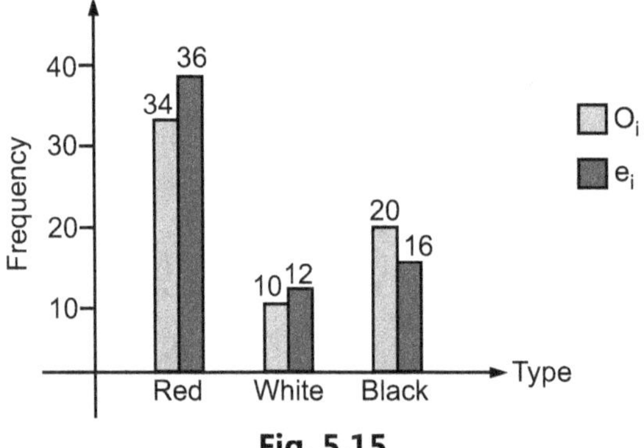

Fig. 5.15

Example 5.14 : One hundred samples were drawn from a production process each after 5 hours. The number of defectives in these sample were noted. A Poisson distribution by estimating the parameter m was fitted to these data. The results obtained are as follows :

Number of detectives	Number of samples (observed)	Expected number of samples
0	63	60.65
1	28	30.33
2	6	7.58
3	2	1.26
4	1	0.16
5 and above	0	0.02

Test the goodness of fit of Poisson distribution in above situation.

[Use 5 % level of significance]

Solution : We want to test H_0 : Fitting of Poisson distribution is good (proper).

Against H_1 : Fitting of Poisson distribution is not proper.

Here we pool expected frequencies until their sum becomes ≥ 5 and also pool corresponding observed frequencies. Thus the frequencies can be written as :

Observed frequencies (o_i)	Expected frequencies (e_i)
63	60.65
28	30.33
9	9.02

We use the test statistic

$$\chi^2_{k-p-1} = \sum_{i=1}^{k} \left(\frac{o_i^2}{e_i}\right) - N$$

Here number of parameters estimated $= p = 1$, $N = 100$, $k = 3$

$$\therefore \qquad \chi^2_1 = 100.27009 - 100$$

$$\therefore \qquad \chi^2_1 = 0.27009 \qquad\qquad \text{[Calculated value]}$$

$$< \chi^2_{1;\,0.05} = 3.841 \qquad\qquad \text{[Table value]}$$

Hence we accept H_0 at 5 % l.o.s.

Another approach. Also p-value $= P[\chi^2_1 > 0.27009] = 0.6032 > 0.05$. Accept H_0.

Conclusion : Fitting of Poisson distribution may be good to the given data.

5.2 Test Based on F-Distribution

Test for equality of population variances

In some situations we require to test that the observations in two groups have same variability. Also, it is the basic assumption for testing equality of population means using t-test.

Let $X_1, X_2, \dots X_i, \dots X_{n_1}$ be a random sample of size n_1 (< 30) from normal population with mean μ_1 and variance σ_1^2. Similarly $Y_1, Y_2, \dots Y_i, \dots Y_{n_2}$ denote the observations in a random sample of size n_2 drawn from another the population with mean μ_2 and variance σ_2^2. We desire to test $H_0 : \sigma_1^2 = \sigma_2^2$.

Then mean square for the first sample

$$= s_1^2 = \frac{\sum (x_i - \bar{x})^2}{n_1 - 1} = \frac{\sum x_i^2 - n_1 \bar{x}^2}{n_1 - 1}$$

Similarly mean square for the second sample

$$= s_2^2 = \frac{\sum (y_i - \bar{y})^2}{n_2 - 1} = \frac{\sum y_i^2 - n_1 \bar{y}^2}{n_1 - 1}$$

In this case sampling distribution of

$$U = \frac{(n_1 - 1)\, s_1^2}{\sigma_1^2} \text{ is } \chi^2 \text{ with } (n_1 - 1) \text{ degrees of freedom}$$

Also, $$V = \frac{(n_2 - 1)\, s_2^2}{\sigma_2^2} \text{ follows } \chi^2 \text{ distribution with } (n_2 - 1) \text{ degrees of}$$

freedom

$\because$ Samples are independent, U and V are independent

$$\therefore \quad F = \frac{\dfrac{U}{n_1 - 1}}{\dfrac{V}{n_2 - 1}} = \frac{\dfrac{s_1^2}{\sigma_1^2}}{\dfrac{s_2^2}{\sigma_2^2}} \rightarrow \text{F-distribution with } (n_1 - 1) \text{ and } (n_2 - 1) \text{ degree of freedom.}$$

Under $H_0 : \sigma_1^2 = \sigma_2^2,$ $F = \dfrac{s_1^2}{s_2^2} \rightarrow$ F-distribution with $(n_1 - 1)$ and $(n_2 - 1)$ d.f.

Since, the values of F-distribution given in statistical tables are constructed for F ratio ≥ 1, hence larger out of s_1^2, s_2^2 is to be taken in the numerator of the test statistic and the first degree of freedom are the degrees of freedom of r.v. in numerator and the second degrees of freedom are that of the r.v. in denominator.

Thus, if $s_1^2 > s_2^2$ Under $H_0 : \sigma_1^2 = \sigma_2^2$ the statistic.

$$F = \frac{s_1^2}{s_2^2} \text{ has F-distribution with } (n_1 - 1) \text{ and } (n_2 - 1) \text{ d.f.}$$

On the other hand if $s_2^2 > s_1^2$, under H_0 the statistic $F = \dfrac{s_2^2}{s_1^2}$ follows F-distribution with $(n_2 - 1)$ and $(n_1 - 1)$ d.f.

Hence if $s_1^2 > s_2^2$ the critical region at level of significance α will be shaded region as shown below :

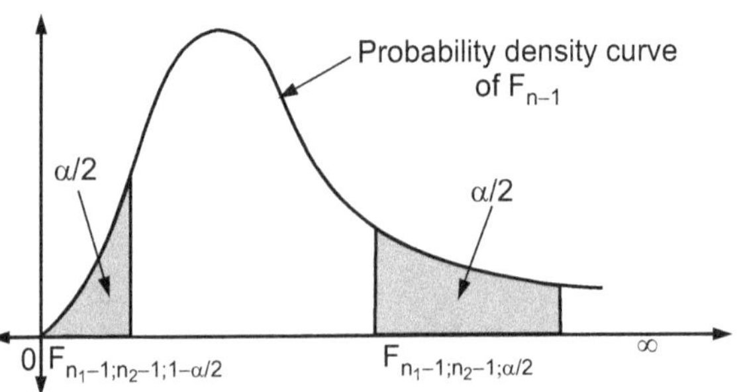

Fig. 5.16

Thus we reject H_0 at level of significance α if,

$$F_{n_1-1,\, n_2-1} = \frac{s_1^2}{s_2^2} \geq F_{n_1-1,\, n_2-1;\, \frac{\alpha}{2}}$$

$$\text{Or } F_{n_1-1,\, n_2-1} = \frac{s_1^2}{s_2^2} \leq F_{n_1-1,\, n_2-1;\, 1-\frac{\alpha}{2}}$$

and accept H_0 otherwise. On the other hand when $s_2^2 > s_1^2$ we reject H_0 at *l.o.s.* α if

$$F_{n_2-1,\, n_1-1} = \frac{s_2^2}{s_1^2} \geq F_{n_2-1,\, n_1-1;\, \frac{\alpha}{2}}$$

$$\text{Or } F_{n_2-1,\, n_1-1} = \frac{s_2^2}{s_1^2} \leq F_{n_2-1,\, n_1-1;\, 1-\frac{\alpha}{2}}$$

Remark : The value of $F_{n_1-1,\, n_2-1;\, 1-\alpha/2}$ can be obtained by taking reciprocal of $F_{n_2-1,\, n_1-1;\, \alpha/2}$ In other words

$$F_{n_1-1,\, n_2-1;\, 1-\alpha/2} = \frac{1}{F_{n_2-1,\, n_1-1;\, \alpha/2}}$$

Justification : From Fig. 5.9, we have,

$$P\left(F < F_{n_1-1,\, n_2-1;\, 1-\alpha/2}\right) = \frac{\alpha}{2}$$

$$\therefore \quad P\left[\frac{1}{F} > \frac{1}{F_{n_1-1,\, n_2-1;\, 1-\alpha/2}}\right] = \frac{\alpha}{2}$$

$$\therefore \quad P\left[n_2-1,\, n_1-1 > \frac{1}{F_{n_1-1,\, n_2-1;\, 1-\alpha/2}}\right] = \frac{\alpha}{2} \qquad \ldots \text{(i)}$$

Now, $\quad P\left[F_{n_2-1,\, n_1-1} > F_{n_2-1,\, n_1-1;\, \alpha/2}\right] = \frac{\alpha}{2} \qquad \ldots \text{(ii)}$

Comparing (i) and (ii),

$$\frac{1}{F_{n_1-1,\, n_2-1;\, 1-\alpha/2}} = F_{n_2-1,\, n_1-1;\, \alpha/2}$$

Thus, $\quad F_{n_1-1,\, n_2-1;\, 1-\alpha/2} = \frac{1}{F_{n_2-1,\, n_1-1;\, \alpha/2}}$

Note :

(1) The values of $F_{n_1-1,\, n_2-1;\, \alpha}$ are available in the statistical tables for $\alpha = 0.1$ (5 percent points of e^{2Z}) and for $\alpha = 0.02$ (1% point of e^{2Z}).

(2) We can get judgement about $\sigma_1 = \sigma_2$ from box plot.

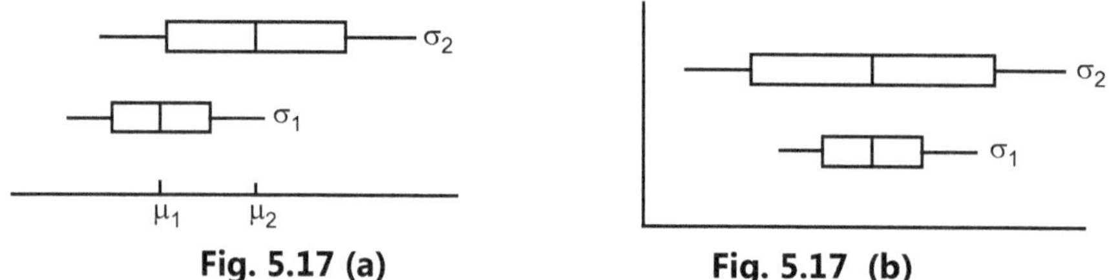

Fig. 5.17 (a)   **Fig. 5.17 (b)**

Further if we construct box plot for $X_i - \mu_1$ and $Y_i - \mu_2$ the centres of boxes get aligned.

Example 5.15 : Two manufacturers A and B supply piston rods of specified diameters to a company. The company is interested in comparing variability of the diameters of the product of these two manufacturers. The measurements (in cm) of the diameters of the rods drawn randomly from the rods supplied by A and B are as follows.

[Use 2 % level of significance]

Diameters of rods supplied by A : 6.4, 6.8, 7.2, 6.6, 7.1, 7.0, 6.5, 7.1, 6.9, 7.3

Diameters of rods supplied by B : 6.9, 6.5, 6.6, 7.4, 6.2, 6.8, 7.0, 7.3, 7.1, 7.2, 6.4, 6.7

Test whether variances in the diameters of rods supplied by manufacturers A and B are equal. Using (A) Statistical tables, (B) MS-Excel, (C) R software.

Solution : (A) Let, σ_1^2 : Variance in the diameters of the product of the manufacturer A.

σ_2^2 : Variance in the diameters of the product of the manufacturer B.

We want to test $H_0 : \sigma_1^2 = \sigma_2^2$ against $H_1 : \sigma_1^2 \neq \sigma_2^2$

Now s_1^2 = Mean square of the sample drawn from first population.

$$= \frac{\Sigma x_i^2 - n_1 \, \bar{x}^2}{n_1 - 1} = \frac{475.57 - (10)\,(6.89)^2}{9}$$

$$= \frac{475.57 - 474.721}{9} = 0.094333$$

$$s_2^2 = \frac{\Sigma y_i^2 - n_2 \, \bar{y}^2}{n_2 - 1}$$

$$= \frac{563.25 - (12)\,(6.84167)^2}{11}$$

$$= \frac{1.54863}{11} = 0.1407845$$

$\because s_2^2 > s_1^2$, we take test statistic as

$$F_{11,\,9} = \frac{s_2^2}{s_1^2} = 1.4924204 \qquad \text{[Calculated value]}$$

$$< F_{11,\,9;\,0.02} = 5.11 \qquad \text{[Table value]}$$

$\therefore$ We accept H_0 at 2 % l.o.s.

Second method using p-value approach :

P $[F_{11, 9} > 1.4924]$ = 0.2786 = p-value > level of significance = 0.02. $\therefore$ Accept H_0 at 2% level of significance.

Conclusion : The variances in the diameters of rods supplied by two manufacturers A and B may be equal.

(B)   Using MS-EXCEL :

The above problem can be solved using ME-EXCEL for one tailed alternative at 5% level of significance for one tailed alternative at 5% level of significance as follows :

Solution : Step 1 : Take observations of sample I in some column and observations of sample II in another column as follows :

	A	B
1	6.4	6.9
2	6.8	6.5
3	7.2	6.6
4	6.6	7.4
5	7.1	6.2
6	7	6.8
7	6.5	7
8	7.1	7.3
9	6.9	7.1
10	7.3	7.2
11		6.4
12		6.7

Fig. 5.18 (a)

Step 2 : Click on Tools and then Data analysis .

Step 3 : Select F-test two sample for variances .

Step 4 : The box given below appears on the screen.

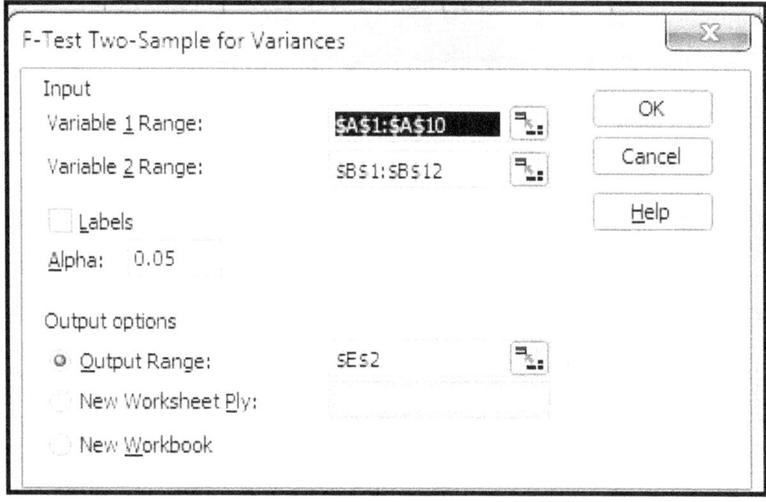

Fig. 5.18 (b)

Step 5 : Click on OK then output will be as follows :

F-Test Two-Sample for Variances		
	Variable 1	Variable 2
Mean	6.89	6.84166667
Variance	0.094333333	0.14083333
Observations	10	12
df	9	11
F	0.669822485	
P(F<=f) one-tail	0.278455192	
F Critical one-tail	0.322322225	

Fig. 5.18 (c)

While reading the output we have to keep in mind that in MS-EXCEL F test is carried out for one tailed test only.

Here P [F $\leq$ 0.669822485] $\leq$ 0.278455 which is P value [P [F $\leq$ f) gives p value].

$\therefore$ This P vlaue > level of significance = 0.05.

$\therefore$ Accept H_0 at 5% level of significance more details about F test to be carried out using MS-EXCEl are as follows :

The tool calculates the value f of an F-statistic (or F-ratio). A value of f close to 1 provides evidence that the underlying population variances are equal. In the output table, if f < 1 "P (F <= f) one-tail" gives the probability of observing a value of the F-statistic less than f when population variances are equal and "F critical one-tail" gives the critical values less than 1 for the chosen significance level, Alpha. If f > 1, "P(F <= f) one-tail" gives the probability of observing a value of the F-statistic greater than f when population variances are equal and "F Critical one-tail" gives the critical value greater than 1 for Alpha.

(c) Solution using R software :

```
>x=c (6.5, 6.8, 7.2, 6.6, 7.1, 7.0, 6.5, 7.1, 6.9, 7.3)
>y=c (6.9, 6.5, 6.6, 7.4, 6.2, 6.8, 7.0, 7.3, 7.1, 7.2, 6.4, 6.7)
>var.test(x, y)
```

data: x and y

F= 0.6698, num df = 0, denom df = 11, p-value = 0.5569

alternative hypothesis : true ratio of variance is not equal to 1

95 percent confidence interval :

0.1866894   2.6203954
sample estimates:
ratio of variances
0.6698225

Points to Remember

(A)   Tests based on χ^2 distribution :

(i) χ^2 test for independence of two attributes A and B arranged in 2×2 contingency table. In this test H_0 : Two attributes A and B are independent.

A \ B	B_1	B_2	Total
A_1	a	b	a + b
A_2	c	d	c + d
Total	a + c	b + d	N = a + b + c + d

The test statistic is

$$\chi^2 = \frac{(ad - bc)^2 \, N}{(a + b)(c + d)(a + c)(b + d)} \to \chi^2_1$$

which follow χ^2 distribution with 1 d.f.

Remark : If any cell frequency is less than 5, we use following corrected test statistic

$$\chi^2_1 = \frac{\left\{ |ad - bc| - \dfrac{N}{2} \right\}^2}{(a + b)(c + d) + (a + c)(b + d)}$$

This correction is called Yate's correction.

(ii) χ^2 test for independence of two attributes arranged in $r \times s$ contingency table. In this case test statistic is

$$\chi^2 = \sum_{i=1}^{r} \sum_{j=1}^{s} \frac{(O_{ij} - e_{ij})^2}{e_{ij}} = \sum_{i=1}^{r} \sum_{j=1}^{s} \left(\frac{O_{ij}^2}{e_{ij}} \right) - N$$

where N = Total of observed frequencies.

It has χ^2 distribution with $(r - 1)(s - 1)$ d.f.

(iii)   McNemar's test : To test H_0 of symmetry for 2×2 contingency table. Suppose 2×2 contingency table is as follows :

A \ B	B_1	B_2
A_1	O_{11}	O_{12}
A_2	O_{21}	O_{22}

Alternatively test statistic is given by

$$H_0 \ : \ P_{1\bullet} = P_{\bullet 1} \text{ against } H_1 : P_{1\bullet} \neq P_{\bullet 1}$$

OR

$$H_0 \ : \ P_{2\bullet} = P_{\bullet 2} \text{ against } H_1 : P_{2\bullet} \neq P_{\bullet 2}$$

The test statistic is

$$\chi^2_1 = \frac{(O_{12} - O_{21})^2}{O_{12} + O_{21}} \text{ where } O_{11}, O_{12}, O_{21}, O_{22} \text{ are observed frequencies.}$$

(iv) χ^2 Test for goodness of fit :

In this test null hypothesis is

$$H_0 \; : \; \text{Fitting of a probability distribution is good}$$

OR

$$H_0 \; : \; \text{There is no significant difference between observed and expected frequencies.}$$

In this case the test statistics is

$$\chi^2 \; = \; \sum_{i=1}^{k} \frac{(O_i - e_i)^2}{e_i} \; = \; \sum_{i=1}^{k} \left(\frac{O_i^2}{e_i} \right) - N$$

where o_i's are observed frequencies, e_i's are expected frequencies and $N = \sum_{i=1}^{n} O_i$. It has χ^2 distribution with $k - p - 1$ d.f. where, p = number of parameters estimated. k = number of frequencies after pooling.

Assumption : $e_i \geq 5$.

Remark : When expected frequency of a class is less than 5, the class is merged into neighbouring class along with its observed and expected total of expected frequencies becomes ≥ 5.

(v)

(i) Test for $H_0 : \sigma_2 = \sigma_0^2$ where σ^2 = population variance when mean is known.

In this test, the test statistic is $\dfrac{\sum_{i=1}^{n} (X_i - \mu)^2}{\sigma_0^2}$. It follows χ^2 distribution with n d.f.

(ii) Test for $H_0 : \sigma^2 = \sigma_0^2$ where σ^2 is population variance when mean is unknown.

In this test statistic $\dfrac{\sum_{i=1}^{n} (X_i - \bar{X})^2}{\sigma_0^2}$ follows χ^2 distribution with $(n - 1)$ d.f.

(B) Tests based on t-distribution :

(1) One sample t test :

In this test $H_0 : \mu = \mu_0$ and population variance σ^2 is unknown.

Under $H_0 : \mu = \mu_0,$ $\; t = \dfrac{\bar{X} - \mu_0}{s/\sqrt{n}} = \dfrac{(\bar{X} - \mu_0)\sqrt{n}}{s}$ where $s^2 = \dfrac{\sum (X_i - \bar{X})^2}{n-1} = \dfrac{\sum X_i^2 - n\bar{X}^2}{n-1}$ follows t distribution with $(n - 1)$ d.f.

$\therefore$ The two sided confidence interval [C.I.] for μ (when σ is unknown) with confidence coefficient $(1 - \alpha)$ is $\left(\bar{X} - t_{n-1;\, \alpha/2} \dfrac{s}{\sqrt{n}} , \; \bar{X} + t_{n-1;\, \alpha/2} \dfrac{s}{\sqrt{n}} \right).$

(ii) Two-sample t test : In order to test $H_0 : \mu_1 = \mu_2$ when population variances are unknown but equal this test is used. Under $H_0 : \mu_1 = \mu_2$ [i.e. equality of population means].

$$\frac{\bar{X} - \bar{Y}}{s\sqrt{\dfrac{1}{n_1} + \dfrac{1}{n_2}}} \text{ follows t distribution with } n_1 + n_2 - 2 \text{ d.f.}$$

where

$$s^2 = \frac{1}{n_1 + n_2 - 1} [\Sigma (X_i - \bar{X})^2 + (Y_i - \bar{Y})^2]$$

$$= \frac{1}{n_1 + n_2 - 1} [(\Sigma X_i^2 - n_1 \bar{X}^2) + (\Sigma Y_i^2 - n_2 \bar{Y}^2)]$$

$\therefore$ The two sided confidence interval for difference between two population means with confidence coefficient $(1 - \alpha)$ is given by

$$\left[(\bar{X} - \bar{Y}) - t_{n_1 + n_2 - 2;\, \alpha/2}\, s\sqrt{\frac{1}{n_1} + \frac{1}{n_2}}\,,\, (\bar{X} - \bar{Y}) + t_{n_1 + n_2 - 2;\, \alpha/2}\, s\sqrt{\frac{1}{n_1} + \frac{1}{n_2}} \right]$$

(iii) Paired t-test :

When observations in the populations can be paired as before and after this test is used to test $H_0 : \mu_d = 0$ (Average difference is 0). Suppose $X_1 \ldots X_i \ldots X_n$ are observations before and $Y_1, Y_2, \ldots, Y_i \ldots Y_n$ are observations after then order $H_d : \mu_d = 0$.

$$t = \frac{\bar{d}}{s/\sqrt{n}} = \frac{\bar{d}\sqrt{n}}{s} \text{ has t distribution with } (n - 1) \text{ d.f.}$$

where

$$\bar{d} = \frac{\Sigma d_i}{n}$$

and

$$s^2 = \frac{\Sigma (d_i - \bar{d})^2}{n - 1} = \frac{\Sigma d_i^2 - n \bar{d}^2}{n - 1}$$

(C) Test based on F distribution :

Let $X_1, X_2 \ldots X_i \ldots X_{n_1}$ be a random sample from $N(\mu_1, \sigma_1^2)$ distribution and $Y_1, Y_2 \ldots Y_i \ldots Y_{n_2}$ be a random sample from $N(\mu_2, \sigma_2^2)$. Then order $H_0 : \sigma_1^2 = \sigma_2^2$ the statistics.

$$\frac{s_1^2}{s_2^2} \longrightarrow F_{n_1 - 1,\, n_2 - 1}$$

where

$$s_1^2 = \frac{\Sigma (X_i - \bar{X})^2}{n_1 - 1} = \frac{\Sigma X_i^2 - n_1 \bar{X}^2}{n_1 - 1}$$

and $$s_2^2 = \frac{\sum (Y_i - \bar{Y})^2}{n_2 - 1} = \frac{\sum Y_i^2 - n_2 \bar{Y}^2}{n^2 - 1}$$

Also $\dfrac{s_2^2}{s_1^2} \longrightarrow F_{n_2 - 1,\, n_1 - 1}$

Exercise 5 (A)

1. Explain the test procedure for testing the independence of two attributes in a $r \times s$ contingency table. **(P.U. April 98, 99, Oct. 98)**

2. Explain the procedure for carrying out the test for testing $H_0 : \mu_1 = \mu_2$ against :

    (i) $H_1 : \mu_1 \neq \mu_2$

    (ii) $H_1 \cdot \mu_1 > \mu_2$

    (iii) $H_1 : \mu_1 < \mu_2$ for small samples only. **(P.U. April 2000)**

3. Explain the test procedure for testing $H_0 : \sigma_1^2 = \sigma_2^2$ against $H_1 : \sigma_1^2 \neq \sigma_2^2$. **(P.U. Oct. 2011, 2012)**

4. State situations in which paired t-test can be used. **(P.U. April 98)**

5. Explain the small sample test for testing the null hypothesis that population mean has specified value μ_0.

6. Explain in detail paired t-test along with the assumptions made. Give one illustration in which this test can be used. **(Oct. 98)**

7. Describe χ^2-test for goodness of fit. State the assumptions we make while applying the test. **(P.U. May 2012)**

8. State the procedure for testing null hypothesis that population variance has specified value σ_0^2.

9. How will you test, the independence of two attributes each at 2 levels ?

10. Explain a small sample test for testing the agreement between observed and expected frequencies.

11. Distinguish between two sample t-test for testing equality of two population means and paired t-test.

12. State situations in which a test based on χ^2-distribution can be used.

13. (a) State situations in which a two sample test based on t-distribution can be used.

    (b) With usual notations, state the test statistic and critical region for testing $H_0 : \sigma_1^2 = \sigma_2^2$ against $H_1 : \sigma_1^2 < \sigma_2^2$. State the underlying assumptions. **(May 2001)**

14. Explain the term $(1 - \alpha)$ 100% confidence interval for population parameter.

15. State $100\,(1 - \alpha)\,\%$ confidence interval for

    (i) Population mean μ when population S.D. is known.

    (ii) Population mean μ when population S.D. is unknown.

    (iii) Difference between means of two normal populations.

16. Write a short on McNemar's test.

17. Describe the test procedure used to test symmetry of contingency table in McNemar's test.

18. Explain the test procedure for testing equally of two population means when population variance are equal but unknown. Also state 95% confidence interval for difference in population means. **(P.U. May 2013)**

Exercise 5 (B)

1. The table below gives the number of accidents that occurred in the certain factory on the various days of a particular week.

Days of week	Sun.	Mon.	Tues.	Wed.	Thurs.	Fri.	Sat.
No. of accidents	6	4	9	7	8	10	12

Test at 5 % level whether the accidents are uniformly distributed over the different days.

2. The following is a 2 × 2 contingency table.

Eye colour in father	Eye colour in son	
	Not light	Light
Not light	23	15
Light	15	47

Test whether the eye colour in son is associated with the eye colour in father.
[use α = 0.05] **[P.U. April 2000]**

3. In a sample of 8 observations the sum of squares of deviations of the sample values from the sample mean was 84. In another sample of 10 observations it was 99. Test whether the difference between population means is significant at 5 % level.

4. A die when tossed 300 times gave the following results :

Score	1	2	3	4	5	6
Frequency	43	49	56	45	66	41

Are the data consistent at 5 % level of significance with the hypothesis that the die is true ?

5. A random sample of size 15 from a bivariate normal population gave a correlation coefficient of 0.3. At 2 % level of significance check whether the value of population correlation coefficient is significant.

6. Theory predicts that the proportion of beans in 3 groups A, B, and C should be in the ratio 1 : 2 : 3. In an experiment on 300 beans the frequencies in the 3 groups were found to be 45, 105 and 150 respectively. Does the experiment support theory ? Justify.

7. The table below gives the number of books issued from a certain library on the various days of a week.

Days	Mon.	Tues.	Wed.	Thurs.	Fri.	Sat.
No. of books issued	120	130	110	115	135	110

Test at 5 % l.o.s. whether the issuing of books is independent of a day.

8. In a locality 100 persons were randomly selected and asked for their educational achievements. The results are given as under.

Sex	Education		
	Primary school	High school	College
Male	10	15	25
Female	25	10	15

Test whether education depends on sex at 1 % level of significance.

9. A random sample of 10 boys had the following I Q'S

70, 120, 110, 101, 88, 83, 95, 88, 107, 100.

(i) Does these data support the assumption that population mean I.Q. is 100 ?

[use α = 0.05)

(ii) Construct 95% confidence interval for mean I.Q. of boys. Also find length of C.I.

10. In an experiment on pea breeding, a scientist obtained the following frequencies of seeds : 316 round and yellow, 102 wrinkled and yellow, 109 round and green and 33 wrinkled and green. Theory predicts that the frequencies of seeds should be in the proportion 9 : 3 : 3 : 1 respectively. Set a proper hypothesis and test it at 5 % l.o.s.

11. Test a hypothesis H_0, an experiment was performed three times and the resulting values of chi-square are 2.37, 2.86 and 3.54 each of which corresponding to 1 d.f. Show that H_0 cannot be rejected at 5 % l.o.s. on the basis of any individual experiment but it is rejected when the results of the these experiments are pooled together.

12. A certain stimulus is administered to each of 12 patients resulted in the following increase in blood pressure (b.p.)

5, 2, 8, –1, 3, 0, –2, 1, 5, 0, 4 and 6. Can it be concluded that the administration of the stimulus in general will be accompanied by increase in the b.p. ? (use α = 0.05)

13. Two random samples size 9 and 11 d.f. are drawn from two normal populations. The following information is given :

Sample No.	Sample size	ΣX_i	ΣX_i^2
I	9	9.6	61.52
II	11	16.5	73.26

Test whether two populations have same variance [use α = 0.05]

14. Selfed progenics of a cross between pure strains of plants were segregated as follow :

	Early flowering : β	Late flowering : β
Tall : A	120	48
Short : α	36	12

Do these results agree with the assumption that the cell frequencies in ratio (AB) : (αB) : (Aβ) : ($\alpha\beta$) : : 9 : 3 : 3 : 1 ?   [Use α = 0.05] **[P.U. April 99)**

15. The following information is collected on two characters :

	Cinegoers	Non-cinegoers
Literate	83	57
Illiterate	45	65

Based on this information can you concluded that there is no association between habit of watching cinema going and literacy.   [Use α = 0.05]

16. A company has installed new colour video display terminals to replace the old ones. A batch of 22 operators were trained to use new machine. They required on the average 17.2 hours with variance of 35.8 squared hours for achieving satisfactory performance. In order to learn old terminals the company had trained 20 operators for on the average 21.3 hours with standard deviation of 7 hours. Should the supervisors of the company claim that new terminals are easier to operate ?   [Take α = 0.01]

17. Ten computer dealers in different areas of the city were asked the prices of dot matrix printers (in thousands of Rs.) of two companies A and B. The prices recorded are given below :

Dealer	Company A	Company B
1	17	20
2	18	19
3	16	17
4	18.5	18
5	17.8	18.5
6	16.9	17.2
7	17.6	19
8	16.6	17.5
9	20	21
10	18.6	19.3

Is it reasonable to assume that on the average the printer of company A is cheaper than company B ? [Take α = 0.01]

18. A news paper publisher is interested in testing whether newspaper readership in the society is associated with readers' educational achievement. A related survey showed the following results :

Type of readership	Level of Education			
	Post graduate	Graduate	Passed S. S. C	Not passed S. S. C.
Never	09	12	30	60
Sometimes	25	20	15	20
Daily	68	48	40	10

Test whether type of newspaper readership depends on level of education.

[Take α = 0.05]

19. A bank found that the variance in the number of days between account transactions for pass book is 80 days squared. After implementing a new policy that penalizes the customer with a service charge, a sample of 22 saving accounts was drawn. It showed variance between the transaction as 26 days squared. Is the bank justified in claiming that the new policy reduces the variances of days between transactions ?

[Use 1 % *l.o.s.*]

20. From the information given below test whether the type of occupation and attitude towards the social laws are independent   [Use 1 % *l.o.s.*]

	Attitude towards social laws		
Occupation	**Favourable**	**Neutral**	**Opposite**
Blue-collar	29	26	37
White collar	25	32	56
Professional	34	21	42

21. Certain pesticide is packed into bags by a machine. A random sample of 10 bags is drawn and their weights (in kg) are found as follows : 50, 49, 52, 44, 45, 48, 46, 45, 49, 45.

(i) Test if the average weight of a bag can be taken as 50 kg ?

(ii) Construct 99% confidence interval for mean weight of bag.

22. Life expectancy in 10 regions of India in 1950 and 12 regions of India in 1980 is as follows :

Regions	**Life expectancy (in years)**	
	1950	**1980**
1	37	44
2	39	45
3	36	47
4	35	43
5	44	42
6	45	50
7	50	52
8	41	48
9	44	51
10	42	43
11	-	46
12	-	49

Test whether variance of life expectancy in 1950 is same as that in 1980 [Use 2 % l.o.s.]

23. A man power development officer is asked to determine whether weekly wages of unskilled workers in the two cities A and B are same. He obtained following results from the sample of workers selected from each city.

City	**Number of workers in the sample**	**Mean weekly wages (in Rs.)**	**Variance of wages**
A	9	600	121
B	8	640	144

Using these data should he conclude that there is significant difference in the weekly wages in the two cities ?

24. The following table shows the classification of 1200 workers in a factory according to the disciplinary action taken by the management and their promotional experience.

Disciplinary action	Promotional experience	
	Promoted	Not promoted
Non-offenders	100	258
Offenders	42	800

Test whether the promotional experience is independent of deciplinary action.

25. A farmer utilises fertiliser of company A on a 7 acre farm and fertiliser of company B on a 9 acre farm to harvest rice. After measuring the yield, he found yield per acre as 13 quintals with standard deviation of 4.3 quintals on the farm on which fertiliser of company A was applied. The corresponding values for the farm on which fertiliser of company B was applied are 15 quintals and 5.2 quintals respectively. Do these data indicate that the two fertilisers are equally effective under the assumption that all conditions like temperature, supply of water, soil fertility, method of cultivation etc. were identical ?

26. Manager of a hotel from his past experience knows that 75 % of customers order tea, 20 % order Coffee and 5 % order milk. After long period of time, in a random sample of 500 customers, 400 ordered tea, 90 ordered coffee, 400 ordered tea, 90 ordered coffee and 10 ordered milk. Should the manager conclude that the drinking habits of customers have changed over the period of time ?

27. The following table shows the opinions of men and women about the legality of abortion. Test whether there is significant difference between means and womens opions.

	Favourable	Opposed	Undecided
Men	2	28	20
Women	8	30	12

28. A publisher claims that average expenditure on text books in a year of students in Senior College of University is Rs. 150. In order to test the claim 28 students were selected at random. It was observed that these students spent on an average Rs. 140 on the books with standard deviation of Rs. 35. Will you conclude that the claim of publisher is true from the results of survey ?

29. Test whether the following sample can be regarded as a sample drawn from a normal population with variance 4. Use 5% level of significance.

7.95, 4.90, 3.65, 7.07, 4.60.

30. The weight of 15 bags of salt taken from a machine are found are follows :

Weight (in kg) 15.8, 16.1, 16.1, 16.0 16.0, 15.8, 15.7, 16.2, 15.9, 15.7, 15.7, 15.8, 16.0, 16.0, 15.8.

(i)   Does the sample fail in supporting the hypothesis that the machine gives bag with weight of 16.0 kg on an average. Use 5% level of significance.

(ii) Obtain 95% confidence interval for mean weight of bag. Also find length of C.I.

31.   In an experiment the effect of a pesticide, "Dimecron – 100" was tudied in relation to mortality of fish "Tilapia mossambica" in two different aquaria. The pesticide was used at two levels; low concentration and high concentration in both the aquaria. The following observations were recorded.

Types of the dose	No. of dead fish	
	Aquarium 1	Aquarium 2
Low concentration	27	39
High concentration	73	61

Test at 5% level of significance whether the type of dose of the pesticides has any effect on the mortality of the fish.

32.   The following table gives weights (in grams) of 8 albino rats in two different batches.

Weights in first batch : 105, 95, 85, 102, 108, 103, 95, 87.

Weight in second batch : 110, 101, 90, 110, 112, 107, 98, 91.

Test whether the two populations from which the two batches were selected have equal variances. Use 10% l.o.s.

33.   A standardised placement test in mathematics was given to 25 boys and 13 girls. The boys made an average grade of 82 with standard deviation of 8, while the girls made an average grade of 78 with standard deviation of 7. Test the hypothesis $H_0 : \sigma_1^2 = \sigma_2^2$ against $H_1 : \sigma_1^2 \neq \sigma_2^2$ at 2% level of significance.

34.   From the following data, test whether defective production is associated with shift at 5% level of significance.

Type of Unit	Shift		
	Day	Evening	Night
Defective units	10	5	20
Non-defective units	30	20	20

35.   Below are given the gain in weights (in/lbs) of pigs fed on two diets A and B.

Diet	Gain in weight
A	25, 32, 30, 34, 24, 14, 32, 24, 30, 31, 35, 25
B	44, 34, 22, 10, 47, 31, 40, 30, 32, 35, 18, 21, 35, 29, 22.

(i) Test whether the two diets differ significantly as regards to their effect on increase in weight.

(ii) Consturct 95% confidence interval for difference between mean gains in weights of pigs due to two diets A and B. Also find length of C.I.

36.   The following table gives the grades in a certain examination to students of different high schools according to their locations.

Grades Location	Fail	Pass	Second-class	First-class
City	24	80	65	28
Town	17	48	35	17
Village	9	42	30	15

Test whether the grades are associated with location of high schools. Use 5% l.o.s.

37. Ten specimens of copper wires drawn from a large lot have the following breaking strength (in kg. wt) 578, 572, 570, 568, 578, 570, 569, 548, 572, 572. (i) Test whether the population mean strength from which the lot is selected is 578 kg wt. at 5% level of significance. (ii) Mean breaking strength of copper wires also find length of C.I.

38. The following are the values (in thousands of an inch) obtained by two engineers in 10 successive measurements with the same micrometer.

Engineer A : 503, 505, 497, 505, 495, 502, 499, 495, 510, 501.

Engineer B : 502, 497, 492, 498, 499, 495, 497, 496, 498, 500.

Test the hypothesis $H_0 : \sigma_A^2 = \sigma_B^2$ against the alternative $H_1 : \sigma_A^2 \neq \sigma_B^2$ at 10% level of significance.

39. The prices of shares (in Rs.) of Titan watches observed on 10 different days in a month were as follows : 66, 65, 69, 70, 69, 71, 70, 63, 64, 68.

(i)   Test whether the mean price of share in a month in Rs. 65. Use 5% level of significance.

(ii)   Obtain 95% confidence interval for mean price of shares.

40. The time taken by workers in performing a job by method I and method II is given below :

Method I : 20, 16, 26, 27, 23, 22.

Method II : 27, 33, 42, 35, 32, 34, 38.

Do the data show that the variances of time distribution of populations from which these samples are drawn do not differ significantly ? Use 10% l.o.s.

41. Two random samples drawn from two normal populations are given below :

Sample I : 20, 16, 26, 27, 23, 22, 18, 24, 25, 19.

Sample II : 17, 23, 32, 25, 22, 24, 28, 31, 33, 20, 22.

Test whether the two populations have the same variance. Use 2% level of significance.

42. The following table shows the distribution of tall, medium, short men married to all medium and short women in a sample of 700 couples.

Wife Husband	Short	Medium	Tall
Short	20	60	60
Medium	90	160	100
Tall	90	80	40

Do these data suggest that the height has any effect on choice of the partner at the time of marriage. (Use 5% l.o.s.)

43. The hours of sleep for 10 patients before and after giving a new drug and recorded. Test whether there is a significant difference in the average hours of sleep at 5% level of significance.

Patient No.	1	2	3	4	5	6	7	8	9	10
Hours of sleep (before)	6	5	7	7	8	9	6	6	7	8
Hours of sleep (after)	7	8	7.5	9	7	6.5	7	8	8.5	7

44. The following table shows the classification of persons according to residential area and respiratory diseases.

Residential area	Respiratory disease	
	Present	**Absent**
Farm	20	130
Around factory	40	60

Test whether respiratory disease depend upon residential area. Use 5% level of significance. **(P.U. May 2013)**

45. The following table gives information about colour blindness of right and left eye of a group of people in certain city.

		Colour blindness in left eye	
		Present	**Absent**
Colour blindness in right eye	Present	20	41
	Absent	39	400

Test whether vision quality of two eyes is same as far as colour blindness is concerned.

[Use 5% *l.o.s.*]

46. The 2×2 contingency table for two attributes A and B is as follows :

A \ B	B_1	B_2
A_1	50	35
A_2	6	9

Test whether the two attributes A and B are equally effective. [Use 5% *l.o.s.*]

Exercise 5 (C)

Choose correct alternative out of (a) to (d) in the following question.

1. Let X_1, X_2, ... X_{12} be a random sample from a normal population with mean μ and *unknown* variance σ^2. Then under $H_0 : \mu = 15$, the statistic $\dfrac{\left(\bar{X} - 15\right)\sqrt{12}}{s}$ where s^2 is sample mean square follows :

    (a)   t with 12 d.f.
    (b)   t-distribution with 15 d.f.
    (c)   t-distribution with 14 d.f.
    (d)   t-distribution with 11 d.f.

2. Suppose $X_1, X_2, \ldots X_{10}$ is a random sample from normal population with mean μ and *known* variance σ^2. Then under $H_0 : \mu = 4$, the statistic $\left(\dfrac{\overline{X} - 4\sqrt{10}}{\sigma}\right)$ follows :

(a) t-distribution with 9 d.f.
(b) t-distribution with 10 d.f.
(c) N (0, 1) distribution
(d) t-distribution with 4 d.f.

3. Let $X_1, X_2, \ldots X_{17}$ be a random sample from a normal population with mean μ and *unknown* variance σ^2. Suppose sample mean square is denoted by s^2. Then which of the following is the appropriate test statistic used to test $H_0 : \mu = 15$.

(a) $\dfrac{\left(\overline{X} - 15\right)\sqrt{16}}{s}$ (b) $\dfrac{\left(\overline{X} - 15\right)\sqrt{17}}{s}$ (c) $\dfrac{\left(\overline{X} - 15\right)\sqrt{15}}{s}$ (d) $\dfrac{\left(\overline{X} - 16\right)\sqrt{17}}{s}$

4. Let $X_1, X_2, \ldots X_9$ be a random sample from normal population with mean μ and unknown variance σ^2. We want to test $H_0 : \mu = 16$ against $H_1 : \mu \neq 16$ at level of significance α. In this case, the critical region is given by calculated valued of the appropriate test statistic is –

(a) $< t_{8;\ \alpha/2}$ (b) $\geq t_{8;\ \alpha/2}$ (c) $\geq t_{9;\ \alpha/2}$ (d) $< t_{9;\ \alpha/2}$

5. Suppose $X_1, X_2, \ldots X_{10}$ is a random sample from N (μ_1, σ^2) population while $Y_1, Y_2, \ldots Y_{12}$ is a random sample from N (μ_2, σ^2) population where σ^2 is unknown. The mean square of the pooled sample comes out to be 64. Then in order to test $H_0 : \mu_1 = \mu_2$, the appropriate test statistic will be –

(a) $\dfrac{\overline{X} - \overline{Y}}{8\sqrt{\dfrac{1}{9} + \dfrac{1}{11}}}$ (b) $\dfrac{\overline{X} - \overline{Y}}{64\sqrt{\dfrac{1}{10} + \dfrac{1}{12}}}$ (c) $\dfrac{\overline{X} - \overline{Y}}{64\sqrt{\dfrac{1}{9} + \dfrac{1}{11}}}$ (d) $\dfrac{\overline{X} - \overline{Y}}{8\sqrt{\dfrac{1}{10} + \dfrac{1}{11}}}$

6. Paired t-test was applied using 13 pairs observations $\{(X_i, Y_i); i = 1, 2, \ldots 13\}$ where x_i : observation before applying a method and y_i : observation corresponding to x_i after applying the method. In this case, the distribution of test statistic under the null hypothesis $H_0 : \mu_a = 0$ is –

(a) t-distribution with 13 d.f.   (b) t-distribution with 12 d.f.

(c) t-distribution with 26 d.f.   (d) t-distribution with 24 d.f.

7. Suppose a sample of size 27 is drawn from a bivariate normal population to test $H_0 : \rho = 0$ where ρ is population correlation coefficient. It gives the sample correlation coefficient between two variance as 0.6. Then calculated value of the appropriate test statistic to test H_0 is –

(a) $\dfrac{30}{8}$ (b) $\dfrac{\sqrt{30}}{8}$ (c) $\dfrac{8}{30}$ (d) $\dfrac{30}{\sqrt{8}}$

8. Suppose a sample $\{(x_i, y_i); i = 1, 2, ...n; n < 30\}$ is drawn from a bivariate normal population to test $H_0 : \rho = 0$ where ρ denotes population correlation coefficient between X and Y. It gives the value of sample correlation coefficient as 'r'. Then the appropriate test statistic to test H_0 is :

(a) $\dfrac{r\sqrt{n-1}}{\sqrt{1-r^2}}$ (b) $\dfrac{r\sqrt{n-2}}{\sqrt{1-r^2}}$ (c) $\dfrac{r\sqrt{n-2}}{\sqrt{1-r}}$ (d) $\dfrac{\sqrt{r}\sqrt{n-1}}{\sqrt{1-r^2}}$

9. Let $X_1, X_2, ... X_{15}$ be a random sample from the normal population with mean 20 and variance σ^2. We want to test $H_0 : \sigma^2 = 25$. The value of $\sum\limits_{i=1}^{15} (X_i - 20)^2$ comes out to be 150.

Then the calculated value of appropriate test statistics to test H_0 is

(a) 10 (b) 15 (C) 30 (D) 6

10. Suppose $X_1, X_2, ... X_{21}$ be a random sample from the normal population with known mean μ and unknown variance σ^2. Then to test $H_0 : \sigma^2 = \sigma_0^2$ the appropriate test statistic is,

(a) $\dfrac{\sum\limits_{i=1}^{21} \left(X_i - \bar{X}\right)^2}{\sigma_0^2}$ (b) $\dfrac{\sum\limits_{i=1}^{21} (X_i - \mu)^2}{(20)\,\sigma_0^2}$

(c) $\dfrac{\sum\limits_{i=1}^{21} (X_i - \mu)^2}{\sigma_0^2}$ (d) $\dfrac{\sum\limits_{i=1}^{21} (X_i - \mu)^2}{(21)\,\sigma_0^2}$

11. Suppose $X_1, X_2, ... X_{15}$ be a random sample from a normal population with unknown parameters μ and σ^2. Then to test $H_0 : \sigma^2 = \sigma_0^2$ the appropriate test statistic is,

(a) $\dfrac{\sum\limits_{i=1}^{15} (X_i - \mu)^2}{\sigma_0^2}$ (b) $\dfrac{\sum\limits_{i=1}^{15} \left(X_i - \bar{X}\right)^2}{\sigma_0^2}$

(c) $\dfrac{\sum\limits_{i=1}^{15} \left(X_i - \bar{X}\right)^2}{\sigma_0^2}$ (d) $\dfrac{\sum\limits_{i=1}^{15} (X_i - \mu)^2}{15\,\sigma_2^0}$

12. Suppose $O_1, O_2, ... O_i, ... O_k$ is a set of observed frequencies and $e_1, e_2, ... e_i, ... e_k$ are corresponding and $\sum\limits_{i=1}^{K} o_i = N$. Then in order to test $H_0 :$ There is no significant between observed and expected frequencies the test statistic used is

(a) $\displaystyle\sum_{i=1}^{k}\left(\frac{e_i}{o_i^2}\right) - N$

(b) $\displaystyle\sum_{i=1}^{k}\left(\frac{o_i^2}{e_i}\right) - N$

(c) $\displaystyle\sum_{i=1}^{k}\left(\frac{e_i}{o_i}\right) - N$

(d) $\displaystyle\sum_{i=1}^{k}\left(\frac{e_1^2}{o_i}\right) - N$

13. Suppose $e_1, e_2, ... e_i, ... e_{10}$ ($e_i \geq 5$ for all i) is a set of expected frequencies obtained after fitting a probability distribution in which 2 parameters were estimated. Then under
H_0 : Fitting of the probability distribution is good, the test statistic used follows :

(a) χ^2-distribution with 5 d.f.

(b) χ^2-distribution with 9 d.f.

(c) χ^2-distribution with 7 d.f.

(d) χ^2-distribution with 10 d.f.

14. A 4×3 contingency table was obtained to test H_0 : Two attributes A and B are independent. Then under H_0, the distribution of test statistic used in the test.

(a) χ^2 with 6 d.f.

(b) χ^2 with 12 d.f.

(c) χ^2 with 7 d.f.

(d) χ^2 with 11 d.f.

15. We want to test H_0 : Two attributes A and B are independent and both the attributes are at two levels. Then under H_0 the statistic used in this case follows :

(a) χ_2^2

(b) χ_4^2

(c) χ_3^2

(d) χ_1^2

16. Suppose a random sample $X_1, X_2, ... X_{12}$ drawn from normal population with parameters μ_1 and σ_1^2 has mean square s_1^2 while a random sample $Y_1, Y_2, ... Y_{10}$ drawn from another normal population with parameters μ_2 and σ_2^2 has mean square s_2^2. Hence, under

$H_0 : \sigma_1^2 = \sigma_2^2$, the statistic $\dfrac{s_1^2}{s_2^2}$ follows :

(a) F-distribution with 9 and 11 d.f.

(b) F-distribution with 11 and 9 d.f.

(c) F-distribution with 12 and 10 d.f.

(d) F-distribution with 10 and 12 d.f.

Exercise 5 (D)

State whether the following statements are true or false.

1. A t-test cannot be applied only if the observations in the population from which random sample is drawn follow normal distribution.

2. In a paired t-test observations in two samples are independent of each other.

(P.U. May 2011)

3. A random sample X_1, X_2, ... X_{25} drawn from normal population with unknown parameters μ and σ^2 has mean square 625. Then under $H_0 : \mu = 50$, the statistic $\dfrac{\overline{X} - 50}{5}$ follows t-distribution with 24 d.f.

4. Suppose X_1, X_2, ... X_{10} is a random sample from normal population with mean μ_1 and variance σ_1^2. Another random sample Y_1, Y_2, ... Y_9 is drawn from normal population with mean μ_2 and variance σ_2^2 where $\sigma_2^2 \neq \sigma_1^2$. Then in order to test $H_0 : \mu_1 = \mu_2$. we can use the test statistic $\dfrac{\overline{X}\,\overline{Y}}{s\sqrt{\dfrac{1}{10} + \dfrac{1}{9}}}$ where s^2 = mean square of pooled sample.

5. In a test based on t-distribution, the value of the test statistic cannot be negative.

6. In a test based on χ^2-distribution we need not pool frequencies of neighbouring classes while computing the value of test statistic if any expected frequency is less than 5.

7. The critical region in χ^2-test of goodness of fit is always one sided.

8. Suppose A and B are two attributes each at 2 levels. We want to test H_0 : A and B are independent. In this case we use test statistic which follows chi-square distribution with 1 d.f. under H_0.

9. Suppose X_1, X_2, ... X_{14} is a random sample from normal population with known mean μ and unknown variance σ^2. Then to test $H_0 : \sigma^2 = 50$ the appropriate test statistic is
$$\dfrac{\sum\limits_{i=1}^{14} \left(X_i - \overline{X}\right)^2}{50}.$$

10. The two attributes A and B are at levels 5 and 4 respectively. We want to test H_0 : A and B are independent. All cell frequencies are greater than 5. Then the statistic we use in the test follow χ^2-distribution with 12 d.f. under H_0 (True)

11. Chi-square test for goodness of fit is always right tailed test.

12. For using F-test in order to test H_0 : Variances of two normal population are equal, it is necessary that means of two population should be equal.

13. MC Neymar's test is used to test the hypothesis of symmetry of 2×2 contingency table.

Answers of Exercise 5 (B)

1. $\chi_6^2 = 5.25$ $\chi_{6;\,0.05}^2 = 15.592$ Accept H_0.

2. $\chi_1^2 = 13.2$ $\chi_{1;\,0.05}^2 = 3.841$ Reject H_0.

3. $t_{16} = -0.6242407$ $t_{16;\,0.05} = 2.12$ Accept H_0.

4. $\chi_5^2 = 8.56$ $\chi_{5;\,0.05}^2 = 11.07$ Accept H_0.

5. $t_{12} = 1.0894994$ $t_{12;\,0.01} = 3.055$ Accept H_0.

6. $\chi_2^2 = 0.75$, $\chi_{2;\,0.05}^2 = 5.991$ Accept H_0.

7. $\chi_5^2 = 4.583333$ $\chi_{5;\,0.05}^2 = 11.07$ Accept H_0.

8. $\chi_2^2 = 9.929$ $\chi_{2;\,0.05}^2 = 5.991$ Accept H_0.

9. $t_9 = -0.83$ $t_{9;\,0.05} = 2.262$,   Accept H_0,
    p value = 0.4305,   95% C.I. is (85.785, 106.616)

10. $\chi_3^2 = 0.3555554$, $\chi_{3;\,0.05}^2 = 7.815$ Accept H_0.

12. $t_{11} = 2.90$, $t_{11;\,0.05} = 2.201$ Reject H_0

13. $F_{8,\,10} = 1.3378684$, $F_{8;\,10;\,0.05} = 3.07$ Accept H_0.

14. $\chi_3^2 = 2.1574073$, $\chi_{3;\,0.05}^2 = 7.815$ Accept H_0.

15. $\chi_1^2 = 8.326$, $\chi_{1;\,0.05}^2 = 3.841$ Reject H_0.

16. $t_{40} = -1.9962903$, $t_{40;\,0.01} = 2.021$ Accept H_0.

17. $t_{17} = -1.74$, $t_{17;\,0.01} = 2.11$,   Accept H_0.

18. $\chi_6^2 = 97.651$ $\chi_{6;\,0.05}^2 = 12.592$ Reject H_0.

19. $\chi_{21}^2 = 7.15$, $\chi_{21;\,0.01}^2 = 32.671$ Accept H_0.

20. $\chi_4^2 = 5.415$ $\chi_{4;\,0.01}^2 = 9.488$ Accept H_0.

21. $t_9 = -3.2$, $t_{9;\,0.05} = 2.262$ Reject H_0.
    p value = 0.01084   99% C.I. is (44.558, 50.043)

22. $F_{9,\,11} = 1.9250607$, $F_{9;\,11;\,0.01} = 4.65$ Accept H_0.

23. $t_{15} = -6.73$, $t_{15;\,0.05} = 2.26$ Reject H_0 at 5 % *l.o.s.*

24. $\chi_1^2 = 126.75696$ $\chi_{1;\,0.05}^2 = 3.841$ Reject H_0 at 5 % *l.o.s.*

25. $t_{14} = -0.1490404$, $t_{14;\,0.05} = 2.145$ Accept H_0

26. $\chi_2^2 = 11.666667$, $\chi_{2;\,0.05}^2 = 5.991$ Reject H_0.

27. $\chi_2^2 = 5.66896$, $\chi_{2;\,0.05}^2 = 5.991$ Accept H_0.

28. $\chi_{27}^2 = -1.484615$, $t_{27;\,0.05} = 2.052$ Accept H_0.

29. $\chi_4^2 = 28.46031$, $\chi_{4;\,0.05}^2 = 9.488$ Reject H_0.

30. $t_{14} = -2.2255$ $t_{14;\,0.05} = 2.145$ Reject H_0.
    p-value = 0.04299   95% C.I. is (15.817, 15.997)

31. $\chi_1^2 = 3.256$, $\chi_{1;\,0.05}^2 = 3.841$ Accept H_0.

32. $F_{7,7} = 1.037$, $F_{7,7;\,0.05} = 3.8$ Accept H_0.

33. $F_{24,12} = 1.306$, $F_{24,12;\,0.05} = 2.5$ Accept H_0.

34. $\chi^2_2 = 4.42$ $\chi^2_{2\,;\,0.05} = 5.991$ Accept H_0.

35. $t_{25} = -\,0.6108$ $t_{25;\,0.05} = 2.06$ Accept H_0.

     p-value = 0.5248   95% C.I. is $(-\,8.405,\,4.405)$

36. $\chi^2_6 = 1.627$ $\chi^2_{6\,;\,0.05} = 12.592$ Accept H_0.

37. $t_9 = -\,3.1418$ $t_{9\,;\,0.05} = 2.262$ Reject H_0.

     p-value = 0.01189   99% C.I. is (561.282, 578.518)

38. $F_{9,9} = 1.7505$ $t_{9;\,0.05} = 3.19$ Accept H_0.

39. $t_9 = 2.8246$ $t_{9\,;\,0.05} = 2.262$ Reject H_0.

     p-value = 0.01989   95% C.I. is [65.498, 69.502]

40. $F = 1.1704$ Accept H_0.

41. $F = 4.4222$ Accept H_0.

42. $\chi^2_4 = 4.44$ $\chi^2_{4\,;\,0.05} = 9.488$ Reject H_0.

43. $t_9 = -\,1.22054$, $t_{9;\,0.05} = 1.833$ Accept H_0.

44. $\chi^2_1 = 5.848$ Reject H_0 at 5% *l.o.s.*

45. $\chi^2_1 = 0.05$ Accept H_0 at 1% *l.o.s.*

46. $\chi^2_1 = 20.512$ Reject H_0 at 5% *l.o.s.*

Answers of Exercise 5 (C)

(1)	d	(2)	c	(3)	b	(4)	b
(5)	d	(6)	b	(7)	a	(8)	b
(9)	d	(10)	c	(11)	c	(12)	b
(13)	c	(14)	a	(15)	d	(16)	b

Answers of Exercise 5 (D)

(1)	True	(2)	False	(3)	True	(4)	False
(5)	False	(6)	False	(7)	True	(8)	True
(9)	False	(10)	True	(11)	False.	(12)	False
(13)	True						

APPENDIX

TABLE 1 : THE NORMAL PROBABILITY INTEGRAL

∞		0	1	2	3	4	5	6	7	8	9
0.0	0.	50000	49601	49292	48803	48405	48006	47608	47210	46812	46414
0.1		46017	45620	45224	44828	44433	44038	43644	43251	42858	42465
0.2		42074	41683	41294	40905	40517	40129	39743	39358	38974	38591
0.3		38209	37828	37448	37070	36693	36317	35942	35569	35197	34827
0.4		34458	34090	33724	33360	32997	32636	32276	31918	31561	31207
0.5		30854	30503	30153	29806	29460	29116	28774	28434	28096	27760
0.6		27425	27093	26763	26435	26109	25785	25463	25143	24825	24510
0.7		24196	23885	23576	23270	22965	22663	22363	22065	21770	21476
0.8		21186	20897	20611	20327	20045	19766	19489	19215	18943	18673
0.9		18406	18141	17879	17619	17361	17106	16853	16602	16354	16109
1.0		15866	15625	15386	15151	14917	14686	14457	14231	14007	13786
1.1		13567	13350	13136	12924	12714	12507	12302	12100	11900	11702
1.2		11507	11314	11123	10935	10749	10565	10383	10204	10027	98525
1.3	0.0	96800	95098	93418	91759	90123	88508	86915	85343	83793	82264
1.4		80757	79270	77804	76359	74934	73529	72145	70781	69437	68112
1.5		66807	65522	64255	63008	61780	60571	59380	58208	57053	55917
1.6		54799	53699	52616	51551	50503	49471	48457	47460	46479	45514
1.7		44565	43633	42716	41815	40930	40059	39204	38364	37538	36727
1.8		35930	35148	34380	33625	32884	32157	31443	30742	30054	29379
1.9		28717	28067	27429	26803	26190	25588	24998	24419	23852	23295
2.0		22750	22216	21692	21178	20675	20182	19699	19226	18763	18309
2.1		17864	17429	17003	16586	16177	15778	15386	15003	14629	14262
2.2		13903	13553	13209	12874	12545	12224	11911	11604	11304	11011
2.3		10724	10444	10170	99031	96419	93867	91375	88940	86563	84242
2.4	0.0^2	81975	79763	77603	75494	73436	71428	69469	67557	65691	63872
2.5		62097	60366	58677	57031	55426	53861	52336	50849	49400	47788
2.6		46612	45271	43965	42692	41453	40246	39070	37926	36811	35726
2.7		34670	33642	32641	31667	30720	29798	28901	28028	27179	26354
2.8		25551	24771	24012	23274	22557	21860	21182	20524	19884	19262
2.9		18658	18071	17502	16948	16411	15889	15382	14890	14412	13949
3.0		13499	13062	12639	12228	11829	11442	11067	10703	10350	10008
3.1	0.0^3	96760	93544	90426	87403	84474	81635	78885	76219	73638	71136
3.2		68714	66367	64095	61895	59765	57703	55706	53774	51904	50094
3.3		48342	46648	45009	43423	41889	40406	38971	37584	36243	34946
3.4		33693	32481	31311	30179	29086	28029	27009	26023	25071	24151
3.5		23263	22405	21577	20778	20006	19262	18543	17849	17180	16534
3.6		15911	15310	14730	14171	13632	13112	12611	12128	11662	11213
3.7		10780	10363	99611	95740	92010	88417	84957	81624	78414	75324
3.8	0.0^4	72348	69483	66726	64072	61517	59059	56694	54418	52228	50122
3.9		48096	46148	44274	42473	40741	39076	37475	35936	34458	33037
4.0		31671	30359	29099	27888	26726	25609	24536	23507	22518	21569
4.1		20658	19783	18944	18138	17365	16624	15912	15230	14575	13948
4.2		13346	12769	12215	11685	11176	10689	10221	97736	93447	89337
4.3	0.0^5	85399	81627	78015	74555	71241	68069	65031	62123	59340	56675
4.4		54125	51685	49350	47117	44979	42935	40980	39110	37322	35612
4.5		33977	32414	30920	29492	28127	26823	25577	24386	23249	22162
4.6		21125	20133	19187	18283	17420	16597	15810	15060	14344	13660
4.7		13008	12386	11792	11226	10686	10171	96796	92113	87648	83391
4.8	0.0^6	79333	75465	71779	68267	64920	61731	58693	55799	53043	50418
4.9		47918	45538	43272	41115	39061	37107	35247	33476	31792	30190

(A.1)

Normal Probability Paper

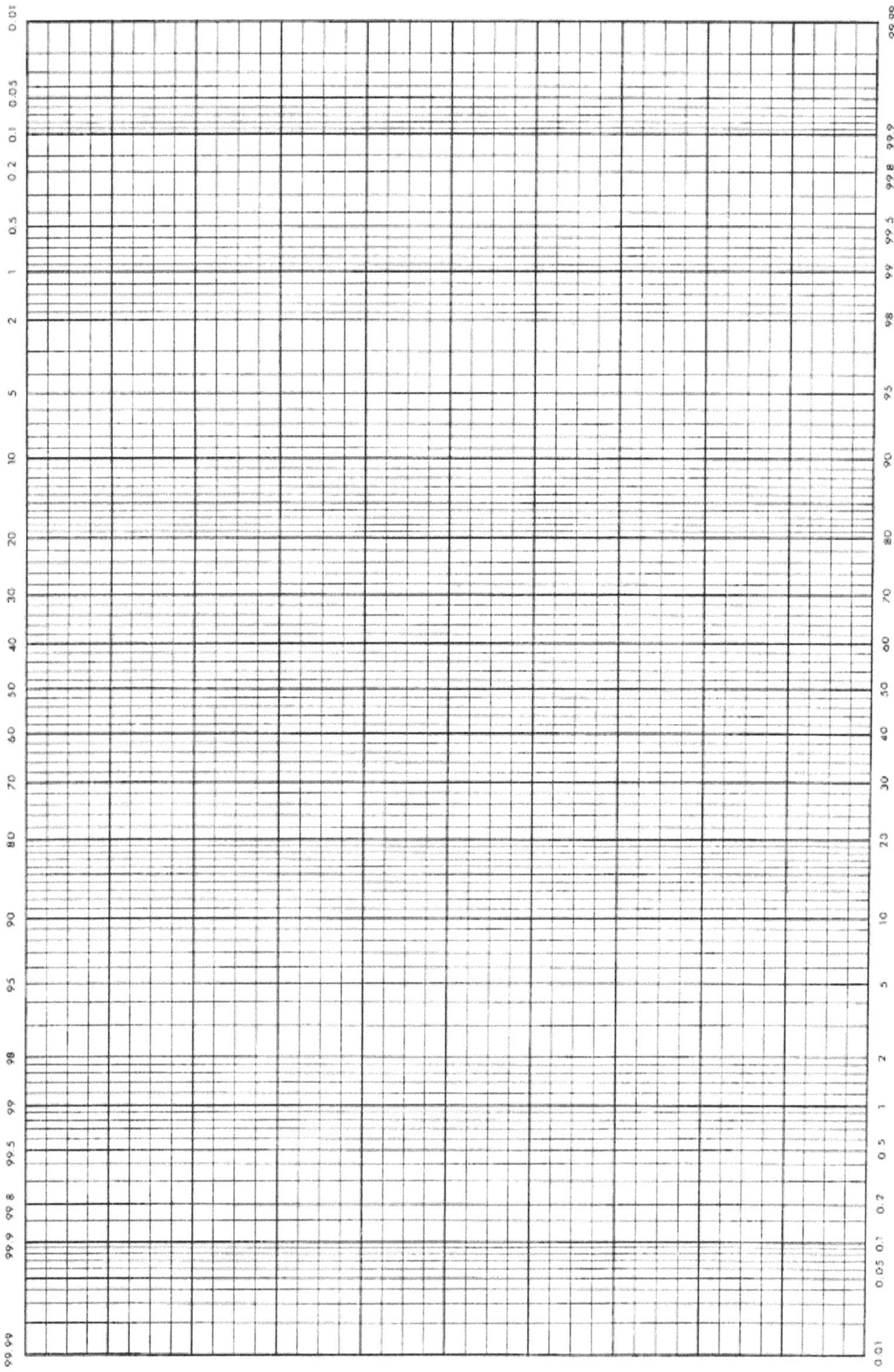

TABLE 2 : DISTRIBUTION OF t
Probability

n	.9	.8	.7	.6	.5	.4	.3	.2	.1	.05	.02	.01	.001
1	158	325	510	.727	1.000	1.376	1.963	3.078	6.314	12.706	31.821	63.657	636.619
2	.142	289	.445	617	816	1.061	1.386	1.886	2.920	4.303	6.965	9.925	31.598
3	.137	.277	.424	.584	.765	.978	1.250	1.886	2.353	3.182	4.541	5.841	12.924
4	.134	.271	.414	.569	.741	.941	1.190	1.533	2.132	2.776	3.747	4.604	8.610
5	.132	267	408	.559	.727	.920	1.156	1.476	2.015	2.571	3.365	4.032	6.869
6	131	265	.404	.553	.718	.906	1.134	1.440	1.943	2.447	3.143	3.707	5.959
7	.130	.263	.402	549	.711	.896	1.119	1.415	1.895	2.365	2.998	3.499	5.408
8	.130	262	399	.546	.706	.889	1.108	1.397	1.860	2.306	2.896	3.355	5.041
9	.129	.261	.398	.543	.703	.883	1.100	1.383	1.833	2.262	2.821	3.250	4.781
10	.129	.260	.397	.542	.700	.879	1.093	1.372	1.812	2.228	2.764	3.169	4.587
11	.129	.260	.396	.540	.697	.876	1.088	1.363	1.796	2.201	2.718	3.106	4.437
12	.128	.259	.395	.539	.695	.873	1.083	1.356	1.782	2.179	2.681	3.055	4.318
13	.128	.259	.394	.538	.694	.870	1.079	1.350	1.771	2.160	2.650	3.012	4.221
14	.128	258	.393	.537	.692	.868	1.076	1.345	1.761	2.145	2.624	2.977	4.140
15	.128	258	393	.536	.691	.866	1.074	1.341	1.753	2.131	2.602	2.947	4.073
16	.128	258	.392	.535	.690	.865	1.071	1.337	1.746	2.120	2.583	2.921	4.015
17	.128	.257	.392	.534	.689	.863	1.069	1.333	1.740	2.110	2.567	2.898	3.965
18	.127	.257	.392	.534	.688	.862	1.067	1.330	1.734	2.101	2.552	2.878	3.922
19	.127	257	.391	.533	.688	861	1.066	1.328	1.729	2.093	2.539	2.861	3.883
20	.127	.257	.391	.533	.687	.860	1.064	1.325	1.725	2.086	2.528	2.845	3.850
21	.127	.257	.391	.532	.686	.859	1.063	1.323	1.721	2.080	2.518	2.831	3.819
22	.127	.256	.390	.532	.686	.858	1.061	1.321	1.717	2.074	2.508	2.819	3.792
23	.127	.256	390	.532	.685	.858	1.060	1.319	1.714	2.069	2.500	2.807	3.767
24	.127	.256	390	.531	.685	.857	1.059	1.318	1.711	2.064	2.492	2.797	3.745
25	.127	.256	390	.531	.684	.856	1.058	1.316	1.708	2.060	2.485	2.787	3.725
26	.127	.256	390	.531	.684	.856	1.058	1.315	1.706	2.056	2.479	2.779	3.707
27	.127	.256	389	.531	.684	.855	1.057	1.314	1.703	2.052	2.473	2.771	3.690
28	.127	.256	389	.530	.683	.855	1.056	1.313	1.701	2.048	2.467	2.763	3.674
29	.127	256	389	.530	.683	.854	1.055	1.311	1.699	2.045	2.462	2.756	3.659
30	.127	.256	389	.530	.683	.854	1.055	1.310	1.697	2.042	2.457	2.75	3.646
40	.126	.255	388	.529	.681	851	1.050	1.303	1.684	2.021	2.423	2.704	3.551
60	.126	.254	.387	.527	.679	.848	1.046	1.296	1.671	2.000	2.390	2.660	3.460
120	.126	.254	.386	.526	.677	.845	1.041	1.289	1.658	1.980	2.358	2.617	3.373
∞	.126	.253	.385	.524	.674	.842	1.036	1.282	1.645	1.960	2.326	2.576	3.291

TABLE 3 : DISTRIBUTION OF χ^2
Probability

n	.99	0.98	.95	.90	.80	.70	.50	.30	.20	.10	.05	.02	.01	.001
1	0.0³157	0.0³628	0.00393	0.0158	0.0642	0.148	0.455	1.074	1.642	2.706	3.841	5.412	6.635	10.827
2	.0201	0.0404	.103	.211	0.446	.713	1.386	2.408	3.219	4.605	5.991	7.824	9.210	13.815
3	0.115	.185	.352	.584	1.005	1.424	2.366	3.665	4.642	6.251	7.815	9.837	11.345	16.266
4	.297	.429	.711	1.064	1.649	2.195	3.357	4.878	5.989	7.779	9.488	11.668	13.277	18.467
5	.554	.752	1.145	1.610	2.343	3.000	4.351	6.064	7.289	9.236	11.070	13.388	15.086	20.515
6	.872	1.134	1.635	2.204	3.070	3.828	5.348	7.231	8.558	10.645	12.592	15.033	16.812	22.457
7	1.239	1.564	2.167	2.833	3.822	4.671	6.346	8.383	9.803	12.017	14.057	16.622	18.475	24.322
8	1.646	2.032	2.733	3.490	4.594	5.527	7.344	9.524	11.003	13.362	15.507	18.168	20.090	26.125
9	2.088	2.532	3.325	4.168	5.380	6.393	8.343	10.656	12.242	14.684	16919	19.679	21.666	27.877
10	2.558	3.059	3.940	4.865	6.179	7.267	9.342	11.781	13.442	15987	18.307	21.161	23.209	29.588
11	3.053	3.609	4.575	5.578	6.989	8.148	10.341	12.899	14.631	17.275	19.675	22.618	24.725	31.264
12	3.571	4.178	5.226	6.304	7.807	9.034	11.340	14.011	15.812	18.549	21.026	24.034	26.217	32.909
13	4.107	4.765	5.892	7.042	8.634	9.926	12.340	15.119	16.985	19.812	22.362	25.472	27.688	34.528
14	4.660	5.368	6.571	7.790	9.467	10.821	13.339	16.222	18.151	21.064	23.685	26.873	29.141	36.123
15	5.229	5.985	7.261	8.547	10.307	11.721	14.339	17.322	19.311	22.307	24.996	28.259	30.578	37.697
16	5.812	6.614	7.962	9.312	11.152	12.624	15.338	18.418	20.465	23.542	26.296	29.633	32.000	39.252
17	6.408	7.255	8.672	10.085	12.002	13.531	16.338	19.511	21.615	24.769	27.587	30.995	33.409	40.790
18	7.015	7.906	9.390	10.865	12.857	14.44	17.338	20.601	22.760	25.989	28.859	32.346	34.805	42.312
19	7.633	8.567	10.117	11.651	13.716	15.352	18.338	21.689	23.900	27.204	30.144	33.687	36.191	43.820
20	8.260	9.237	10.851	12.443	14.578	16.266	19.337	22.775	25.038	28.412	31.410	35.020	37.566	45.315
21	8.897	9.915	11.591	13.240	15.445	17.182	20.337	23.858	26.171	29.615	32.671	36.343	38.932	46.797
22	9.542	10.600	12.338	14.041	16.314	18.101	21.337	24.939	27.301	30.813	33.924	37.659	40.289	48.268
23	10.196	11.293	13.091	14.848	17.187	19.021	22.337	26.018	28.429	32.007	35.172	38.968	41.638	49.728
24	10.856	11.992	13.848	15.659	18.062	19.943	23.337	27.096	29.553	33.196	36.415	40.270	42.980	51.179
25	11.524	12.697	14.611	16.473	18.940	20.867	24.337	28.172	30.675	34.382	37.652	41.566	44.314	52.620
26	12.198	13.409	15.379	17.292	19.820	21.792	25.336	29.246	31.795	35.563	38.885	42.856	45.642	54.052
27	12.879	14.125	16.151	18.114	20.703	22.719	26.336	30.319	32.912	36.741	40.113	44.140	46.963	55.476
28	13.565	14.847	16.928	18.939	21.588	23.647	27.336	31.391	34.027	37.916	41.337	45.419	48.278	56.893
29	14.256	15.574	17.708	19.786	22.475	24.577	28.336	32.461	35.139	39.087	42.557	46.693	49.588	58.302
30	14.953	16.306	18.493	20.599	23.364	25.508	29.336	33.530	36.250	40.256	43.773	47.962	50.892	59.703
32	16.362	17.783	20.072	22.271	25.148	27.373	31.336	35.665	38.466	42.585	46.194	50.487	53.486	62.487
34	17.789	19.275	21.664	23.952	26.938	29.242	33.336	37.795	40.676	44.903	48.602	52.995	56.061	65.247
36	19.233	20.783	23.269	25.643	28.735	31.115	35.336	39.922	42.879	47.212	50.999	55.489	58.619	67.985
38	20.691	22.304	24.884	27.343	30.537	32.992	37.335	42.045	45.076	49.513	53.384	57.969	61.162	70.703
40	22.164	23.838	26.509	29.051	32.345	34.872	39.335	44.165	47.269	51.805	55.759	60.435	63.691	73.402
42	23.650	25.383	28.144	30.765	34.157	36.755	41.335	46.282	49.456	54.090	58.124	62.892	66.206	76.084
44	25.148	26.939	29.787	32.487	35.974	38.641	43.335	48.396	51.639	56.369	60.481	65.337	68.710	78.750
46	26.657	28.504	31.439	34.215	37.795	40.529	45.335	50.507	53.818	58.641	62.830	67.771	71.201	81.400
48	28.177	30.080	33.098	35.949	39.621	42.420	47.335	52.616	55.993	60.907	65.171	70.197	73.683	84.037
50	29.707	31.664	34.764	37.689	41.449	44.313	49.335	54.723	58.164	63.167	67.505	72.613	76.154	86.661
52	31.246	33.256	36.437	39.433	43.281	46.209	51.335	56.827	60.332	65.422	69.832	75.021	78.616	89.272
54	32.793	34.836	38.116	41.183	45.117	48.106	53.335	58.930	62.496	67.673	72.153	77.422	81.069	91.872
56	34.350	36.464	39.801	42.937	46.955	50.005	55.335	61.031	64.658	69.919	74.468	79.815	83.513	94.461
58	35.913	38.078	41.492	44.696	48.797	51.906	57.335	63.129	66.816	72.160	76.778	82.201	85.950	97.039
60	37.485	39.699	43.188	46.459	50.641	53.809	59.335	65.227	68.972	74.397	79.082	84.580	88.379	99.607
62	39.063	41.327	44.889	48.226	52.487	55.714	61.335	67.322	71.125	76.630	81.381	86.953	90.802	102.166
64	40.649	42.960	46.595	49.996	54.336	57.620	63.335	69.416	73.276	78.860	83.675	89.320	93.217	104.716
66	42.240	44.599	48.305	51.770	56.188	59.527	65.335	71.508	75.424	81.085	85.965	91.681	95.626	107.258
68	43.838	46.244	50.020	53.548	58.042	61.436	67.335	73.600	77.571	83.308	88.250	94.037	98.028	109.791
70	45.442	47.893	51.739	55.329	59.898	63.346	69.334	75.689	79.715	85.527	90.531	96.388	100.425	112.317

For odd values of n between 30 and 70 the mean of the tabular values for $n-1$ and $n+1$ may be taken. For larger values of n, the expression $\sqrt{2x^2} - \sqrt{2n-1}$ may be used as a normal deviate with unit variance, remembering that the probability for χ^2 corresponds with that of a single tail of the normal curve. (For fuller formulae, see Introduction.)

TABLE 4 : VARIANCE RATIO—*contd.*
5 per cent. Points of e^{2z}

$$P\,[F_{n1,\,n2} \geq F_{n1,\,n2\,;\,\alpha}] = \alpha$$

$n_2 \backslash n_1$	1	2	3	4	5	6	8	12	24	∞
1	161.4	199.5	215.7	224.6	230.2	234.0	238.9	243.9	249.0	254.3
2	18.51	19.00	19.16	19.25	19.30	19.33	19.37	19.41	19.45	19.50
3	10.13	9.55	9.28	9.12	9.01	8.94	8.84	8.74	8.64	8.53
4	7.71	6.94	6.59	6.39	6.26	6.16	6.04	5.91	5.77	5.63
5	6.61	5.79	5.41	5.19	5.05	4.95	4.82	4.68	4.53	4.36
6	5.99	5.14	4.76	4.53	4.39	4.28	4.15	4.00	3.84	3.67
7	5.59	4.74	4.35	4.12	3.97	3.87	3.73	3.57	3.41	3.23
8	5.32	4.46	4.07	3.84	3.69	3.58	3.44	3.28	3.12	2.93
9	5.12	4.26	3.86	3.63	3.48	3.37	3.23	3.07	2.90	2.71
10	4.96	4.10	3.71	3.48	3.33	3.22	3.07	2.91	2.74	2.54
11	4.84	3.98	3.59	3.36	3.20	3.09	2.95	2.79	2.61	2.40
12	4.75	3.88	3.49	3.26	3.11	3.00	2.85	2.69	2.50	2.30
13	4.67	3.80	3.41	3.18	3.02	2.92	2.77	2.60	2.42	2.21
14	4.60	3.74	3.34	3.11	2.96	2.85	2.70	2.53	2.35	2.13
15	4.54	3.68	3.29	3.06	2.90	2.79	2.64	2.48	2.29	2.07
16	4.49	3.63	3.24	3.01	2.85	2.74	2.59	2.42	2.24	2.01
17	4.45	3.59	3.20	2.96	2.81	2.70	2.55	2.38	2.19	1.96
18	4.41	3.55	3.16	2.93	2.77	2.66	2.51	2.34	2.15	1.92
19	4.38	3.52	3.13	2.90	2.74	2.63	2.48	2.31	2.11	1.88
20	4.35	3.49	3.10	2.87	2.71	2.60	2.45	2.28	2.08	1.84
21	4.32	3.47	3.07	2.84	2.68	2.57	2.42	2.25	2.05	1.81
22	4.30	3.44	3.05	2.82	2.66	2.55	2.40	2.23	2.03	1.78
23	4.28	3.42	3.03	2.80	2.64	2.53	2.38	2.20	2.00	1.76
24	4.26	3.40	3.01	2.78	2.62	2.51	2.36	2.18	1.98	1.73
25	4.24	3.38	2.99	2.76	2.60	2.49	2.34	2.16	1.96	1.71
26	4.22	3.37	2.98	2.74	2.59	2.47	2.32	2.15	1.95	1.69
27	4.21	3.35	2.96	2.73	2.57	2.46	2.30	2.13	1.93	1.67
28	4.20	3.34	2.95	2.71	2.56	2.44	2.29	2.12	1.91	1.65
29	4.18	3.33	2.93	2.70	2.54	2.43	2.28	2.10	1.90	1.64
30	4.17	3.32	2.92	2.69	2.53	2.42	2.27	2.09	1.89	1.62
40	4.08	3.23	2.84	2.61	2.45	2.34	2.18	2.00	1.79	1.51
60	4.00	3.15	2.76	2.52	2.37	2.25	2.10	1.92	1.70	1.39
120	3.92	3.07	2.68	2.45	2.29	2.17	2.02	1.83	1.61	1.25
∞	3.84	2.99	2.60	2.37	2.21	2.10	1.94	1.75	1.52	1.00

Lower 5 per cent. points are found by interchange of n_1 and n_2, i.e. n_1 must always correspond with the greater mean square.

TABLE 5 : VARIANCE RATIO—*contd.*
1 per cent. Points of e^{2z}

$$P\,[F_{n1,\,n2} \geq F_{n1,\,n2\,;\,\alpha}] = \alpha$$

n_2 \ n_1	1	2	3	4	5	6	8	12	24	∞
1	4052	4999	5403	5625	5764	5859	5982	6106	6234	6366
2	98.50	99.00	99.17	99.25	99.30	99.33	99.37	99.42	99.46	99.50
3	34.12	30.82	29.46	28.71	28.24	27.91	27.49	27.05	26.6	26.12
4	21.20	18.00	16.69	15.98	15.52	15.21	14.80	14.37	13.93	13.46
5	16.26	13.27	12.06	11.39	10.97	10.67	10.29	9.89	9.47	9.02
6	13.74	10.92	9.78	9.15	8.75	8.47	8.10	7.72	7.31	6.88
7	12.25	9.55	8.45	7.85	7.46	7.19	6.84	6.47	6.07	5.65
8	11.26	8.65	7.59	7.01	6.63	6.37	6.03	5.67	5.28	4.86
9	10.56	8.02	6.99	6.42	6.06	5.80	5.47	5.11	4.73	4.31
10	10.04	7.56	6.55	5.99	5.64	5.39	5.06	4.71	4.33	3.91
11	9.65	7.20	6.22	5.67	5.32	5.07	4.74	4.40	4.02	3.60
12	9.33	6.93	5.95	5.41	5.06	4.82	4.50	4.16	3.78	3.36
13	9.07	6.70	5.74	5.20	4.86	4.62	4.30	3.96	3.59	3.16
14	8.86	6.51	5.56	5.03	4.69	4.46	4.14	3.80	3.43	3.00
15	8.68	6.36	5.42	4.89	4.56	4.32	4.00	3.67	3.29	2.87
16	8.53	6.23	5.29	4.77	4.44	4.20	3.89	3.55	3.18	2.75
17	8.40	6.11	5.18	4.67	4.34	4.10	3.79	3.45	3.08	2.65
18	8.28	6.01	5.09	4.58	4.25	4.01	3.71	3.37	3.00	2.57
19	8.18	5.93	5.01	4.50	4.17	3.94	3.63	3.30	2.92	2.49
20	8.10	5.85	4.94	4.43	4.10	3.87	3.56	3.23	2.86	2.42
21	8.02	5.78	4.87	4.37	4.04	3.81	3.51	3.17	2.80	2.36
22	7.94	5.72	4.82	4.31	3.99	3.76	3.45	3.12	2.75	2.31
23	7.88	5.66	4.76	4.26	3.94	3.71	3.41	3.07	2.70	2.26
24	7.82	5.61	4.72	4.22	3.90	3.67	3.36	3.03	2.66	2.21
25	7.77	5.57	4.68	4.18	3.86	3.63	3.32	2.99	2.62	2.17
26	7.72	5.53	4.64	4.14	3.82	3.59	3.29	2.96	2.58	2.13
27	7.68	5.49	4.60	4.11	3.78	3.56	3.26	2.93	2.55	2.10
28	7.64	5.45	4.57	4.07	3.75	3.53	3.23	2.90	2.52	2.06
29	7.60	5.42	4.54	4.04	3.73	3.50	3.20	2.87	2.49	2.03
30	7.56	5.39	4.51	4.02	3.70	3.47	3.17	2.84	2.47	2.01
40	7.31	5.18	4.31	3.83	3.51	3.29	2.99	2.66	2.29	1.80
60	7.08	4.08	4.13	3.65	3.34	3.12	2.82	2.50	2.12	1.60
120	6.85	4.70	3.95	3.48	3.17	2.96	2.66	2.34	1.95	1.38
∞	6.64	4.60	3.78	3.32	3.02	2.80	2.51	2.18	1.79	1.00

Lower 1 per cent points are found by interchange of n_1 and n_2, i.e. n_1 must always correspond with the greater mean square.

SPECIMEN QUESTION PAPER

S.Y.B.Sc.

Statistics (Sem. II) (2014 Pattern)

ST 222 : SAMPLING DISTRIBUTIONS AND INFERENCE

Time : 2 Hours   **Maximum Marks : 40**

Instructions to the candidates :

  (1)   All questions are compulsory.

  (2)   Figures to the right indicate full marks.

  (3)   Use of calculator and statistical tables is allowed.

  (4)   Symbols and abbreviations have their usual meanings.

1.   **Attempt each of the following :**   **(1 each)**

  (a)   **Choose the correct alternative in each of the following :**

      (i)   A chi-square random variable has variance 16 then its mode is ......

          (A) 8   (B) 32   (C) 14   (D) 15

      (ii)   If $X_1, X_2, ..., X_{10}$ are i.i.d. $N(0, 1)$ variates then the probability distribution of

$$\frac{3\bar{X}}{\sqrt{\sum\limits_{i=1}^{10} (X_i - \bar{X})^2}} \text{ is } ......$$

          (A) t_{10} (B) χ^2_{10} (C) χ^2_9 (D) t_9

      (iii) if X follows F (6, 8) distribution and Y follows F (8, 6) distribution such that $P(X \geq 9) + P(Y \geq K) = 1$ then value of K is ......

          (A) 9   (B) $\dfrac{1}{9}$ (C) 0.9   (D) 0.1

  (B)   **State whether the following statements are true or false :**   **(1 each)**

      (i)   If X follows F-distribution with (n_1, n_2) d.f. then mean of X is independent of n_2.

      (ii)   In a test based on t-distribution, the value of test statistic cannot be negative.

      (iii) If a random variable follows t-distribution with 2 d.f. then mean deviation about mean is $\sqrt{2}$.

  (c)   State confidence interval for the population mean when population variance is unknown.   **(1)**

  (d)   If X and Y are independent chi-square random variables with 10 and 12 degrees of freedom respectively, identify the probability distribution of $U = \dfrac{12X}{10Y}$.   **(1)**

  (e)   State the rth row moment of chi-square distribution with n degrees of freedom. **(1)**

  (f)   If T follows t-distribution with n d.f. then find $P(|T_n| > 1.812)$.   **(1)**

(S.1)

2. Attempt any two of the following :   **(5 each)**

(a) State and prove additive property of chi-square distribution.

(b) Define student t-distribution. Find $(2r)^{th}$ central moment of t distribution with n d.f. $(2r < n)$.

(c) Let $\bar{X}$ and S^2 be the mean and variance of a random sample of size 25 from $N(3, 100)$ distribution. Compute $P(0 \le \bar{X} \le 6, 55.2 \le S^2 \le 145.6)$.

3. Attempt any two of the following :   **(5 each)**

(a) If X follows t distribution with n d.f. show that X^2 follows F-distribution with 1 and n d.f.

(b) Explain the test procedure for testing equality of two population means when population variances are equal but unknown. Also state 95% confidence interval for difference in population means.

(c) The following table shows the classification of 1200 workers in a factory according to the disciplinary action taken by the management and their promotional experience.

Disciplinary action	Promotional experience	
	Promoted	**Not promoted**
Non-offenders	100	258
Offenders	42	800

Test whether the promotional experience is independent of disciplinary action.

4. Attempt any one of the following :

(a) **(i)** Define Snedecor's F-distribution with n_1 and n_2 d.f. and derive its probability density function (p.f.d.)   **(7)**

  (ii) Write short note on "McNemar's Test".   **(3)**

(b) **(i)** Describe χ^2 test for goodness of fit. State the assumptions we make while applying the test.   **(6)**

  (ii) If a r.v. T follows t-distribution with n degrees of freedom then, show that as $n \to \infty$ the probability distribution of T tends to $N(0, 1)$.   **(4)**

NOV./DEC. 2015

Time : 2 Hours   **Maximum Marks : 40**

Instructions to the candidates :

   (1)   All questions are compulsory.

   (2)   Figures to the right indicate full marks.

   (3)   Use of calculator and statistical tables is allowed.

   (4)   Symbols and abbreviations have their usual meanings.

1. Attempt each of the following :   **(1 each)**

   (a)   **Choose the correct alternative in each of the following :**

        (i)   If a r.v. X follows χ^2 distribution with n d.f. then the variance of X is

           (A) 2n   (B) 4n   (C) n   (D) n (n – 1)

        (ii)   The standard error of a statistic (T) is

           (A) Mean (T)   (B) Median (T)

           (C) Standard Deviation (T)   (D) Variance (T)

        (iii) To test equality of two population means of two Normal Populations with equal unknown σ^2), $H_0 : \mu_1 = \mu_2$ against $H_1 : \mu_1 \neq \mu_2$, the statistic (T) under H_0 follows a

           (A) F distribution with n_1, n_2 degree of freedom

           (B) t distribution with $n_1 + n_2 - 2$ degrees of freedom

           (C) t distribution with n degrees of freedom

           (D) Ch-square distribution with n degrees of freedom

   (B)   **State whether each of the following statements are true or false :**   **(1 each)**

        (i)   A statistic is a function of the parameter values.

        (ii)   If a r.v. T follows t distribution with n degrees of freedom then r.v. T^2 follows F distribution with 1 and n degrees of freedom.

        (iii) $(nS^2)/\sigma^2$ follows a Chi-square distribution with $(n - 1)$ degrees of freedom.

   (c)   What is the value of coefficient of skewness (γ_1) for t distribution with n degrees of freedom ? Give conclusion about the skewness of t distribution.   **(1)**

   (d)   State 100 $(1 - \alpha)$ % confidence interval for population mean μ, when variance σ^2 is unknown.   **(21**

   (e)   State the test statistic of McNemar's test.   **(1)**

   (f)   A r.v. $F \sim F_{n, n}$ distribution. Give the value of its median.   **(1)**

2. Attempt any two of the following :   **(5 each)**

   (a)   State and prove the additive property of χ^2 distribution.

(P.1)

(b) If X_1, X_2, ..., X_n is a random sample from a normal population with mean μ and variance σ^2 (unknown) then explain the procedure to carry out appropriate test for $H_0 : \mu = \mu_0$ against $H_1 : \mu \neq \mu_0$.

(c) If $\bar{x}$ and S^2 are the mean and variance of a random sample of size 25 from $N(\mu = 3, \sigma^2 = 100)$ distribution, evaluate $P\left[(0 \leq \bar{x} < 6) \text{ and } (55.2 < S^2 < 145.6)\right]$.

3. Attempt any two of the following :   (5 each)

(a) If a r.v. T follows Student's t distribution with n degrees of freedom then prove that; as $n \to \infty$, the probability distribution of r.v. T tends to N(0, 1).

(b) If X_1, X_2, ..., X_n is a random sample from a N (μ, σ^2) distribution then explain the test procedure to carry out appropriate test for $H_0 : \sigma^2 = \sigma_0^2$ against $H_1 : \sigma^2 \neq \sigma_0^2$.

(c) The following information is on father's occupation and son's occupation :

	Father in Service	Father in Business
Son in Service	82	18
Son in Business	26	44

Carry out an appropriate test and give conclusions about independence of the two attributes. (use *l.o.s.* $\alpha = 0.05$).

4. Attempt any one of the following :

(a) **(i)** Define Snedecore's F fdistribution. Derive its mean   **(5)**

(ii) A test in English was given to 30 bosy and 20 girls. The average marks for boys is 70 with standard deviation 9 while average marks for girls is 80 with standard deviation 6. Carry out an appropriate test for testing $H_0 : \sigma_1^2 = \sigma_2^2$ against $H_1 : \sigma_1^2 > \sigma_2^2$. (Use *l.o.s.* $\alpha = 0.01$)   **(5)**

(b) **(i)** Explain the paired t-test for testing $H_0 : \mu_d = 0$ against $H_1 : \mu_d \neq 0$. Give a real life situation of it.   **(5)**

(ii) If a r.v. $X \sim F_{m, n}$ distribution and r.v. $Y \sim F_{n, m}$ show that "

$P[X \geq k] + P[Y \geq 1/k] = 1, k > 0$.   **(2)**

(iii) If X_1, X_2 are i.i.d. N (10, 1) random variables then evaluate $P[(X_1 - X_2)^2 < 2.2]$. **(3)**

APRIL 2016

Time : 2 Hours   **Maximum Marks : 40**

Instructions to the candidates :

(1)   All questions are compulsory.

(2)   Figures to the right indicate full marks.

(3)   Use of calculator and statistical tables is allowed.

(4)   Symbols and abbreviations have their usual meanings.

1. Attempt each of the following :

(A)   Choose the correct alternative in each of the following :   **(1 each)**

(a)   If a r.v. X follows χ^2 distribution with n d.f. then mode of the distribution is

    (i)   n + 2   (ii)   n   (iii) n – 2   (iv) 2n

(b)   If a r.v. X follows t-distribution with 5 d.f. then Var (X) is

    (i)   3/5   (ii)   5/3   (iii) 4/3   (iv) 3/4

(c)   If a r.v. X follows $F_{6, n}$ distribution and E(X) = 2 then n is

    (i)   3   (ii)   2   (iii) 6   (iv) 4

(B)   State whether the following statements are true or false :   **(1 each)**

(a)   A statistic is a random variable.

(b)   While testing $H_0 : \sigma^2 = \sigma_0^2$, μ known, based on a random sample of size 12 drawn from a normal population, the test statistic has t_{12} distribution.

(c)   If a r.v. X follows χ_n^2 distribution then $\left(\dfrac{X - n}{\sqrt{2n}}\right)$ follows N(0, 1) distribution.

(C)   State the p.d.f. of F_{n_1, n_2} distribution.   **(1)**

(D)   Define : standard error of a statistic.   **(1)**

(E)   Giev 85% confidence interval for population mean μ when population standard deviation σ is unknown and sample size n is small.   **(1)**

(F)   State the additive property of chi-square distribution.   **(1)**

2. Attempt any two of the following :   **(5 each)**

(a)   Define a χ^2 variate with n d.f. and derive its mean and variance.

(b)   Describe the exact test for testing $H_0 : \mu_1 = \mu_2$ against $H_1 : \mu_1 \neq \mu_2$, where population variances are unknown with equal. $(\sigma_1^2 = \sigma_2^2 = \sigma^2)$.

(c)   The table below gives the number of books issued from a library from Monday to Saturday.

Day	Mon.	Tues.	Wed.	Thu.	Fri.	Sat.
Number of books	120	130	110	115	135	110

Test whether the issuing of books is independent of the day by carrying out an appropriate test. Use 5% *l.o.s.* State your conclusion.

3. Attempt any two of the following :   **(5 each)**

(a) Describe the exact test for testing $H_0 : \sigma_1^2 = \sigma_2^2$ against $H_1 : \sigma_1^2 \neq \sigma_2^2$ when two independent random samples are drawn from normal populations with unknown means.

(b) A health club advertised a weight-reduction and claimed that the average weight (in lbs) reduces after six months program. Check whether the following sample data supports the claim of health club by carrying out an appropriate test. Use 1% l.o.s.

Participant number	1	2	3	4	5	6	7	8
Weight before	120	125	115	130	123	119	122	127
Weight after	111	114	107	120	115	112	112	120

(c) Random variables X_1 and X_2 are i.i.d. $N(0, \sigma^2)$ variates. X_1 and X_2 are transformed to Y_1 and Y_2 by the following orthogonal transformation : $Y_1 = \dfrac{X_1 + X_2}{\sqrt{2}}$ and $Y_2 = \dfrac{X_1 - X_2}{\sqrt{2}}$. Show that Y_1 and Y_2 are independently distributed.

4. Attempt any one of the following :

(a) **(i)** A r.v. Y follows χ_{11}^2 distribution. Obtain

(1) K such that $P\,[Y \leq K] = 0.05$

(2) Median.   **(4)**

(ii) Derive p.d.f. of t distribution n d.f.   **(6)**

(b) **(i)** A r.v. X follows $F_{n_1,\,n_2}$ distribution. Prove that as $n_2 \to \infty$, the distribution of $n_1 X$ tends to $\chi_{n_1}^2$ distribution.   **(5)**

(ii) If $X_1, X_2, \ldots, X_{25}$ is a random sample from $N(5, 9)$ distribution, compute $P\,(3.8 \leq \bar{X} \leq 5.6 \,,\, 5.637 \leq S^2 \leq 11.951)$.   **(5)**

APRIL 2017

Time : 2 Hours   **Maximum Marks : 40**

Instructions to the candidates :

(1) All questions are compulsory.

(2) Figures to the right indicate full marks.

(3) Use of calculator and statistical tables is allowed.

(4) Symbols and abbreviations have their usual meanings.

1. Attempt each of the following :

(A) Choose the correct alternative in each of the following :   (1 each)

    (i) If a random variable (r.v.) X follows chi-square distribution with 8 degrees of freedom (d.f.) then its coefficient of skewness is ......

       (A) $\dfrac{1}{4}$ (B) $\dfrac{1}{2}$ (C) $\dfrac{3}{4}$ (D) $\dfrac{1}{8}$

    (ii) If a r.v. t follows student's t-distribution with g d.f. then the mean of t is ......

       (A) 9   (B) 6   (C) 3   (D) 0

    (iii) If a r.v. F follows Snedecor's F-distribution with 10 and 12 d.f. then mean of Y $= \dfrac{1}{F}$ is ......

       (A) $\dfrac{12}{10}$ (B) $\dfrac{10}{8}$ (C) $\dfrac{8}{10}$ (D) $\dfrac{10}{12}$

(B) State whether the following statements are true or false :   (1 each)

    (i) A statistic is a r.v.

    (ii) If $F \to F_{g \cdot g}$ then median of F = g.

    (iii) If $X \to N(4, 4)$, $Y \to N(4, 5)$ are independent variables then probability distribution of $\dfrac{(X - Y)^2}{g}$ is χ_2^2.

(c) State the test statistic of McNemar's test to test the null hypothesis of equality of marginal distributions.   **(1)**

(d) Define standard error of a test statistic.   **(1)**

(e) If $X_1, X_2, ..., X_{12}$ is a random sample drawn from N(3, 9) distribution then obtain the value of Var $(2\bar{X})$, $\bar{X}$ is sample mean.   **(1)**

(f) If $t \to t_g$ then find constant C such that $P(-C < t < C) = 0.70$.   **(1)**

2. Attempt any two of the following :   (5 each)

(a) If a r.v. X follows chi-square distribution with n d.f. then obtain moment generating function of X. Hence obtain its mean and variance.

(b) Describe the test procedure for testing $H_0 : \mu_1 = \mu_2$, when population variances are unknown but equal, against the following alternatives :

(i) $H_1 : \mu_1 \neq \mu_2$

(ii) $H_1 : \mu_1 < \mu_2$

(iii) $H_1 : \mu_1 > \mu_2$

(c) If $\bar{X}$ and S^2 are the mean and variance of a random sample of size 10 drawn from $N(4, 160)$ distribution, then obtain $P(0 < \bar{X} < 4, 86.08 < S^2 < 170.496)$.

3. Attempt any two of the following :   **(5 each)**

(a) If a r.v. $t \to t_n$, then show that as $n \to \infty$, the probability distribution of r.v. t tends to $N(0, 1)$.

(b) Describe the exact test for testing $H_0 : \sigma_1^2 = \sigma_2^2$, when two independent random samples are drawn from normal populations with unknown means against the following alternatives :

(i) $H_1 : \sigma_1^2 \neq \sigma_2^2$

(ii) $H_1 : \sigma_1^2 > \sigma_2^2$

(iii) $H_1 : \sigma_1^2 < \sigma_2^2$

(c) In a set of random numbers, the digits 0, 1, 2, ... 8, 9 were found to have the following frequencies :

Digit	0	1	2	3	4	5	6	7	8	9
Frequency	43	32	38	27	38	52	36	31	39	24

Test the hypothesis that the digits are uniformly distributed at 5% level of significance.

4. Attempt any one of the following :

(a) **(i)** Define Snedecor's F-distribution. Derive its mean.   **(5)**

(ii) Explain paired t-test for testing $H_0 : \mu_d = 0$ against $H_1 : \mu_d \neq 0$. Give two real life situations of it.   **(5)**

(b) **(i)** Derive the probability density function (p.d.f.) of student's t-distribution with n d.f.   **(6)**

(ii) If a r.v. $X \to \chi_{15}^2$, then find

(I)   Median of X,

(II) Constant K such that $P(X \leq K) = 0.30$.

OCTOBER 2017

Time : 2 Hours   **Maximum Marks : 40**

Instructions to the candidates :
- (1)   All questions are compulsory.
- (2)   Figures to the right indicate full marks.
- (3)   Use of calculator and statistical tables is allowed.
- (4)   Symbols and abbreviations have their usual meanings.

1.   Attempt each of the following :

(A)   Choose the correct alternative in each of the following :   **(1 each)**

(a)   If X and Y are two independent chi-square variates with 10 and 12 degrees of freedom (d.f.) respectively then mode of the distribution is

   (i)   10   (ii)   44   (iii) 22   (iv) 20

(b)   If $X_1, X_2, \ldots, X_{25}$ is a random sample from $N(\mu, \sigma^2)$ and $\sum\limits_{i=1}^{25} (X_i - \bar{X})^2 = 120$ then the distribution of $\sqrt{5}\,(\bar{X} - \mu)$ is

   (i) t_{26} (ii) t_{24} (iii) t_{50} (iv) t_{25}

(c)   Let x_i, $i = 1, 2, \ldots, 6$ be i.i.d. $N(0, 1)$ variates and $U = \dfrac{X_1^2 + X_4^2 + X_3^2}{X_2^2 + X_5^2 + X_6^2}$. Then median of distribution of U is ......

   (i)   3   (ii)   2   (iii) 1   (iv) 0

(B)   State whether the following statements are true or false :   **(1 each)**

(a)   In a test based on t-distribution, the value of the test statistic cannot be negative.

(b)   Let $X_1, X_2, \ldots, X_n$ be a random sample of size n from $N(\mu, \sigma^2)$. To test $H_0 : \mu = \mu_0$ against $H_1 : \mu \neq \mu_0$ when σ is known, the 95% confidence interval for μ is $\left(\bar{X} - 1.96\,\dfrac{\sigma}{\sqrt{n}}, \bar{X} + 1.96\,\dfrac{\sigma}{\sqrt{n}} \right)$.

(c)   If $Y \sim \chi_n^2$ then $\sqrt{2Y} - \sqrt{2n-1}$ follows $N(0, 1)$.

(C)   Explain the term random sample from probability distribution.   **(1)**

(D)   If a continuous random variable X has moment generating function $M_x(t) = (1 - 2t)^{-12}$, identify the distribution of X.   **(1)**

(E)   State a test statistic used to test whether fitting of the particular distribution is good or not.   **(1)**

(F)   State the assumptions to test equality of means of the two populations.   **(1)**

2.   Attempt any two of the following :   **(5 each)**

(a)   Obtain the rth raw moment of chi-square distribution with n d.f. Hence find it's mean.

(b)   State and prove the limiting distribution of a random variable T which follows t-distribution with n d.f.

(c) **(i)** Let X_i, $i = 1, 2, \ldots, 6$ be independent standard normal variates.

Identify the distribution of $U = \dfrac{5X_2^2}{X_1^2 + X_3^2 + X_4^2 + X_5^2 + X_6^2}$.

Hence find the value of 'C' such that $P(U \geq C) = 0.01$.

(ii) Let X be a random variable follows t-distribution with 6 d.f. Find 'a' such that $P(X^2 \leq a) = 0.01$.

3. Attempt any two of the following :   **(5 each)**

(a) If $X_1, X_2, \ldots, X_{n_1}$ is a random sample from $N(\mu, \sigma_1^2)$ distribution and $Y_1, Y_2, \ldots, Y_{n_2}$ is another independent random sample from $N(\mu_2, \sigma_2^2)$ distribution then explain the test procedure for testing $H_0 : \sigma_1^2 = \sigma_2^2$ against $H_1 : \sigma_1^2 \neq \sigma_2^2$.

(b) If $X_1, X_2, \ldots, X_{10}$ is a random sample from $N(4, 9)$ then find
(i) $E(S^2)$ (ii) Var (S^2) (iii) $P(S^2 \geq 17.711)$, where S^2 is a sample variance.

(c) Following are the average weekly losses of work hours due to accidents in 8 industrial plants before and after a certain safety programme was put into operation.

Plant number	1	2	3	4	5	6	7	8
Before	45	73	46	124	33	57	83	34
After	36	60	44	119	35	51	77	29

Test at 5% level of significance (*l.o.s.*) whether the safety programme is effective.

4. Attempt any one of the following :

(a) **(i)** Let X and Y be independent χ^2 variates with m and n d.f. Show that $X + Y$ and $\dfrac{X}{Y}$ are independently distributed.   **(6)**

(ii) The following table gives information about colour blindness of right and left eye of a group of people in certain city.

		Colour blindness in left eye	
		Present	Absent
Colour blindness	Present	100	17
in right eye	Absent	05	8

Test whether vision quality of two eyes is same as far as colour blindness is concerned. Use 5% L.O.S.   **(4)**

(b) **(i)** Let $X_1, X_2, \ldots, X_n$ be a random sample from $N(5, \sigma^2)$. Derive the test statistic to test $H_0 : \sigma^2 = \sigma_0^2$ against $H_1 : \sigma^2 \neq \sigma_0^2$. Give a real life situation of it.   **(5)**

(ii) Define the following terms :
(I)   Statistic (II) Parameter (III) Standard error of statistic   **(3)**

(iii) State mean and mode of F-distribution. Hence discuss the nature of probability curve of it.   **(2)**

◻◻◻

APRIL 2018

Time : 2 Hours   **Maximum Marks : 40**

Instructions to the candidates :
 (1)   All questions are compulsory.
 (2)   Figures to the right indicate full marks.
 (3)   Use of calculator and statistical tables is allowed.
 (4)   Symbols and abbreviations have their usual meanings.

1.   Attempt each of the following :

(A)   Choose the correct alternative in each of the following :   **(1 each)**

   (i)   If X_1 and X_2 are independent random variables (r.v.s) having $N(0, 1)$ and $N(0, 16)$ distribution respectively then probability distribution of $X_1^2 + \frac{1}{4} X_2^2$ is

   (i) $N(0, 17)$ (B) χ_2^2 (C) t_2 (D) $F_{1, 1}$

   (ii) If a r.v. has chi-square distribution with variance equal to 8 then it's moment generating function (m.g.f.) is given by
   (A) $(1 - 2t)^{-2}$ (B) $(1 - t)^{-2}$ (C) $(1 - 2t)^2$ (D) $(1 - t)^2$

   (iii) Suppose e_1, e_2, e_3 and e_4 are expected frequencies such that e_1, $e_2 > 5$ and $e_3 + e_4 = 8$ are obtained after fitting a probability distribution in which one parameter is estimated. Then under H_0 : fitting of probability distribution is good, the test statistics used
   (A) χ^2 with 1 degrees of freedom (d.f.)   (B) χ^2 with 2 d.f.
   (C) χ^2 with 4 d.f.   (D) χ^2 with 3 d.f.

(B)   State whether the following statements is true or false :   **(1 each)**

   (i)   If r.v. F follows $F_{2, 2}$ distribution with $Q_1 = 5$ then $Q_3 = \frac{1}{5}$.

   (ii) Let X be a r.v. having t-distribution with 5 d.f. Then the value of μ_4 is equal to $\frac{25}{2}$.

   (iii) Let X_1, X_2, ..., X_n be a random sample (r.s.) from $N(\mu, \sigma^2)$, μ is unknown. To test $H_0 : \sigma^2 = \sigma_0^2$ against $H_1 : \sigma^2 > \sigma_0^2$ the rejection region is $\chi_{n-1}^2 \leq \chi_{n-1, 1-\alpha}^2$ at α level of significance (l.o.s.).

(c)   State limiting behaviour of χ_n^2 as $n \to \infty$ according to Fisher's approximation.   **(1)**

(d)   State the standard error of the statistics $\dfrac{\sum\limits_{i=1}^{n} (X_i - \bar{X})^2}{n}$.   **(1)**

(e)   Give one real life situation where chi-square test of independence can be used. **(1)**
(f)   Distinguish between two sample t-test and paired t-test.   **(1)**

2. Attempt any two of the following :   **(5 each)**

(a) Find the mode of a chi-square distribution with a d.f. Also if $X \sim \chi_n^2$ and mode of the distribution is 5 find $P(X > 2.167)$.

(b) If a r.v. $U \sim N(0, 1)$, $V \sim \chi_n^2$ and are independent then find the distribution of $\dfrac{U}{\sqrt{\dfrac{V}{n}}}$.

(c) If $X \sim \chi_{10}^2$, $Y \sim \chi_9^2$ and are independent r.v.s then find
   (i) $P[12.242 < Y < 21.666]$
   (ii)   Median of Y
   (iii) $P[X + Y \geq 21.689]$.

3. Attempt any two of the following :   **(5 each)**

(a) If a r.v. $X \sim F_{n_1, n_2}$ then find the distribution of $\dfrac{1}{X}$.

(b) Derive a test statistic to test $H_0 : \sigma_1^2 = \sigma_2^2$ against $H_1 : \sigma_1^2 \neq \sigma_2^2$. Also state the assumptions if any.

(c) If $\bar{X}$ and S^2 are the mean and the variance of a r.s. of size 10 from $N(4, 16)$ then find $P(-1 < X < 4, 6.6688 < S^2 < 17.0496)$.

4. Attempt any one of the following :

(a) **(i)** If $X_1, \ldots, X_n$ is a r.s. from $N(\mu, \sigma^2)$ distribution then show that sample $(\bar{X})$ and sample variance (S^2) are independently distributed.   **(6)**

   (ii) For two independent normal populations we have the following information :

Sample means	$\bar{X} = 10$	$\bar{Y} = 12$
Sample variances	$S_1^2 = 46$	$S_2^2 = 50$
Sample sizes	$n_1 = 15$	$n_2 = 15$

To test $H_0 : \mu_1 = \mu_2$ against $H_1 : \mu_1 \neq \mu_2$, calculate the 95% confidence interval for $(\mu_1 - \mu_2)$. Also give the conclusion. (Use $\alpha = 5\%$)   **(4)**

(b) **(i)** If $X \sim F_{n_1, n_2}$ and $Y \sim F_{n_1, n_1}$ then show that $P(X \geq \alpha) + P\left(Y \geq \dfrac{1}{a}\right) = 1$.   **(2)**

   (ii) A certain stimulus is administrated to each of 12 patients resulted in the following increase in blood pressure : 5, 2, 8, –1, 3, 0, 4, 6, –2, 1, 5, 0.
   Can it be concluded that the administration of the stimulus in general will be accompanied by increase in the b.p. ? Use appropriate test to give the answer. (Use *l.*o.s. = 0.05)   **(4)**

   (iii) Write a short note on McNemar's test.   **(4)**

◻◻◻